The JCT Minor Works Building Contracts 2016

The JCT Minor Works Building Contracts 2016

Fifth Edition

David Chappell
BA(Hons Arch) MA(Arch) MA(Law) PhD RIBA

WILEY Blackwell

This edition first published 2018
© 2018 John Wiley & Sons Ltd

Edition History
Legal Studies and Services (Publishing) Ltd (1e, 1991); Blackwell Science (2e, 1999); Blackwell Publishing (3e, 2004); Blackwell Publishing (4e, 2006)

Registered Offices
John Wiley & Sons, Inc., 111 River Street, Hoboken, NJ 07030, USA
John Wiley & Sons Ltd, The Atrium, Southern Gate, Chichester, West Sussex, PO19 8SQ, UK

Editorial Office
9600 Garsington Road, Oxford, OX4 2DQ, UK

For details of our global editorial offices, customer services, and more information about Wiley products visit us at www.wiley.com.

Wiley also publishes its books in a variety of electronic formats and by print-on-demand. Some content that appears in standard print versions of this book may not be available in other formats.

Library of Congress Cataloging-in-Publication Data

Names: Chappell, David (David M.), author. | Chappell, David (David M.). JCT
   minor works form of contract.
Title: The JCT minor works building contracts 2016 / by Dr. David Chappell,
   BA(Hons Arch), MA(Arch) MA(Law), PhD RIBA David Chappell Consultancy
   Limited, West Yorkshire, United Kingdom.
Description: Fifth edition. | Hoboken : Wiley, 2017. | Revised edition of:
   JCT minor works building contracts 2005 / David Chappell. 4th ed. 2006. |
   Includes bibliographical references and index. |
Identifiers: LCCN 2017016744 (print) | LCCN 2017016886 (ebook) | ISBN
   9781119415282 (pdf) | ISBN 9781119415305 (epub) | ISBN 9781119415541 (pbk.)
Subjects: LCSH: Construction contracts–England.
Classification: LCC KD1615 (ebook) | LCC KD1615 .C48 2017 (print) | DDC 343.4207/869–dc23
LC record available at https://lccn.loc.gov/2017016744

Cover Design: Wiley
Cover Images: (Main Image) © Jose-Luis Saez Martinez / EyeEm/Gettyimages;
(Background Image) © 1xpert/Gettyimages

Set in 10/12.5pt Minion by SPi Global, Pondicherry, India
Printed and bound in Malaysia by Vivar Printing Sdn Bhd

10 9 8 7 6 5 4 3 2 1

Contents

Preface to the fifth edition

The Minor Works Contract continues to be the most popular JCT contract in use. Many architects with whom I have spoken through the RIBA Information Unit have been kind enough to tell me that previous editions of this book have been useful and that they particularly value its straightforward approach. I understand that contractors share that view. Naturally, that is gratifying to know.

I am always trying to produce books which simplify (so far as that is possible!) what the contracts say. In my book about the JCT Standing Building Contract 2011, I tried splitting the subject into more, but shorter, chapters and splitting the chapters into shorter paragraphs with more sub-headings. That proved popular so I have changed the layout of this book accordingly. There are no references to legal cases or clauses in the text in order to make it more readable. The relevant clause numbers are in the margins opposite the text and there are also references to cases and other notes in the margin for those people who like to know more detail.

I have re-written much of the text to make it easier to understand and some explanations from first principles have been included where I thought that would be helpful. I have tried to avoid all legal or pseudo-legal words and I have borne in mind that busy architects and contractors will look into this book to find what to do in various circumstances. Occasionally, where the precise legal position is unclear, I have offered a view. As before, there are some sample letters and some flowcharts to try and make the complicated bits easier to understand.

The 2016 edition of the contract contains a great many changes from the 2011 edition. Many of these changes are apparently minor, but all are significant. The payment provisions have been substantially revised and there are revisions to insurance, variation instructions, CDM Regulations, Supplemental Provisions, definitions, Contractor's Designed Portion and there is scarcely a page without some change.

I have taken the opportunity to remove a few references to legal cases which seemed less important and I have added over 70 cases for those who like to delve more deeply.

I acknowledge the ongoing assistance I receive from Michael Cowlin LLB(Hons) DipOSH Dip Arb FCIArb Barrister (not practising) and Michael Dunn BSc(Hons) LLB LLM FRICS FCIArb who may not know that they have helped to write this book. I wrote the first edition of this book with the eminent contract authority Professor Vincent Powell-Smith. We wrote several books together and I shall always be grateful for his assistance and advice. Finally, I am grateful to my children and grandchildren who brighten my life.

David Chappell
Wakefield
October 2016

Note

1 Throughout the text, the contractor has been referred to as 'it' on the basis that it is a corporate body.
2 The relevant clause or schedule number is shown in italics in the margin opposite the portion of text. Clause numbers referring to MWD have additionally been placed in brackets.
3 Reference numbers to legal cases and other notes are in bold, also in the margin. These notes can be found at the back of the book for those who want them.

Abbreviations

IC   JCT Intermediate Building Contract 2016
ICD   JCT Intermediate Building Contract 2016 with contractor's design
MW   JCT Minor Works Building Contract 2016
MWD JCT Minor Works Building Contract 2016 with contractor's design
SBC   JCT Standard Building Contract With Quantities 2016

1 Introduction

1.1 Some general things about contracts

The law is divided into parts

The construction industry is mostly concerned with the civil law. The civil law governs the way we should behave to our neighbour. We all have rights and duties to each other. They are sometimes set out in Acts of Parliament and sometimes they are derived from the judgments of the courts. Law which is found in the judgments of the courts is usually referred to as the 'common law'. The common law is the system of law which has grown up over the centuries and which is still in the process of evolution. The courts themselves have an order of precedence so that lower courts must follow the decisions of the higher courts. In order to make this work properly, one would imagine that a central system of reporting the judgments of the courts would be in place. In fact, there are a multitude of different organisations producing law reports. Although most of the important judgments are picked up by one or other of these systems, some cases are not reported at all while others are reported by three or four different series of reports. It can be difficult to keep track of the courts' decisions, but this is how the principles of law have been established.

There are also Acts of Parliament such as the Housing Grants, Construction and Regeneration Act 1996 (now amended by a later Act). Often the courts have to rule on the meaning of words in an act of Parliament. Some acts make doing things or failure to do other things a criminal offence for which persons may be fined or imprisoned. Some acts related to the construction industry fall into this category; especially acts dealing with health and safety.

Tort

In general 'tort' is a civil wrong for which the person suffering the wrong is entitled to take action through the courts for compensation. It is based on the duty which everyone owes to one another. There are many wrongs which most people will recognise and which will come under the headings of 'tort'. Most

---

The JCT Minor Works Building Contracts 2016, Fifth Edition. David Chappell.
© 2018 John Wiley & Sons Ltd. Published 2018 by John Wiley & Sons Ltd.

people have heard of negligence, trespass, nuisance and defamation, but there are also some lesser known torts such as interference with contractual rights and breach of statutory duty.

Contract

In addition to legislative or common law rights and duties, two people may agree to have additional rights and duties to each other. For example, I may agree to give Mrs Z advice for a fee of £200. Mrs Z has the right to receive the advice, but a duty to pay me £200 for it. I have a right to the £200, but a duty to provide the advice. If there are agreed rights and duties on both sides, we call it a contract. Of course there may be all kinds of other things which also might have to be agreed, such as the subject matter of the advice, its timing and the date on which Mrs Z must pay me. Even seemingly simple contracts can become quite complicated. If documents have been signed, it is usually said that a contract has been 'executed'. In ordinary language we might say that two people have 'entered into' a contract. Contracts are legally binding which means to say that usually once the contract is agreed, neither person can say: 'I've changed my mind now' without serious consequences.

Breach of contract

If a person does something which the contract does not allow or fails to do something which the contract requires, it is referred to as a 'breach' of contract.

For example, if I do not give the advice which I agreed to give or if the advice is given late or wrong, these are all breaches of contract. The person who is not in breach is usually referred to as the 'injured party' or the 'innocent party'. The injured party is entitled to receive payment from the person in breach to make up for the breach. That is referred to as 'damages'. The amount of money to be paid is normally calculated to put the injured party back in the same position as if the breach had not occurred. Sometimes that is easy, for example, If Mrs Z only pays me £150 for my advice, she could be ordered by a court to pay the additional £50 together with any other costs I had suffered as a result of her failure to pay the full £200 for the advice. Sometimes it is not possible to put someone back in the same position, because the problem is not the shortfall in money but, say, my failure to give proper advice at the right time. The court would try to do what it can to rectify the situation by hearing evidence about what my breach had cost Mrs Z. In such cases the courts have to look into other questions such as to what extent the costs resulting from my breach were reasonably foreseeable at the time we entered into the contract.

Repudiation

If the breach of contract is particularly serious, it may be what is called 'repudiation'. That is a breach which is so serious that it shows that one of the persons does not intend to be bound by the contract any longer. Extreme examples would be if Mrs Z refused to pay anything for my advice or if I refused to give her any advice. In the construction industry, a contractor walking off site, never to return, half way through the project would be repudiation or if the employer told the contractor that he would not be paid any more money.

Faced with repudiation, the injured party has the choice of either accepting the repudiation and seeking damages through the courts, or saying that the contract is still in place and carrying on with it (called 'affirmation'). The injured party is still entitled to seek damages even after affirmation. Obviously, there are many instances where it is just impossible to carry on as if nothing had happened; for example, if the architect stops work half way through preparing construction drawings.

Essentials of a contract

In order for there to be a contract there must be three things:

- Agreement;
- An intention to create legal relations;
- Something given by both persons.

Agreement is usually shown by one person making an offer and another person accepting it. If I offer to give some advice for £200 and Mrs Z accepts we are in agreement.

An intention to create legal relations is usually assumed in commercial dealings and anyone who says that there was no intention has the task of proving it. In a social context, people do not always intend to create legal relationships. If Thomas says to Emma that he will treat her to a meal in a nice restaurant that evening, That is not a contract. If Thomas breaks the arrangement Emma has no redress.

Something given by both persons is fairly straightforward. In the case of my advice I agree to give Mrs Z advice and she gives me £200 in return. In a construction contract, the contractor promises to construct the building and the employer promises to pay whatever is stated in the contract as the Contract Sum. In legal terms, the money or service given is usually referred to as 'consideration'. This consideration can take forms other than the ones just described. For example, one person may agree to pay another, if that second person agrees to stop doing something or not to do something he or she was about to do. The important thing is that both persons contribute something; not necessarily of apparent equal value.

When talking about contracts, it is customary to refer to the 'parties' to the contract. That is convenient when reference to 'persons' would not be appropriate: for example, where one or both parties are corporate bodies such as local authorities, universities or limited companies.

Two types of contract

There are two types of contract:

- Simple contracts;
- Deeds or specialty contracts.

Most contracts are simple contracts. If it is desired to make a contract in the form of a deed, it is necessary to observe a particular procedure. Before 1989, all deeds had to be made by fixing a seal to the document. That could be in wax, but more often it was simply a circular piece of red paper embossed with the name of the relevant party. Nowadays, the procedure is laid down by statute. Essentially, the document must clearly state that it is a deed and the parties must sign in one of the prescribed ways. The alternative ways are usefully set out in JCT contracts on the attestation page.

A deed is a very serious form of contract. Notably:

- There is no need for consideration. In other words, a promise that one party will do something for the other becomes legally binding without any corresponding promise by the other party.
- The limitation period is 12 years (see Chapter 3 below).
- Statements in a deed are conclusive about their truth as between the parties to the deed.

Therefore, it is wise to think very carefully before entering into a contract as a deed.

1.2   Some background to MW and MWD

The JCT Agreement for Minor Works was first published in 1968; it was revised in 1977. The headnote explained that it was intended for minor building works or maintenance work, based on a specification or a specification and drawings, to be carried out for a lump sum. It was inappropriate for use with bills of quantities or a schedule of rates.

Despite its limitations, it was widely used for small projects and even for larger ones, its main attraction being its brevity and apparent simplicity. The evidence suggests that architects were becoming increasingly dissatisfied with the length and complexity of the then current main JCT standard form contract (JCT 63) and then (as now) wished to use simple contract conditions wherever possible.

The Minor Works form was extensively revised by a JCT working party in 1979 and a new edition was published in January 1980 and reprinted with corrections in October 1981. It was revised again in April 1985, January 1987, March and October 1988, September 1989, January 1992 and March 1994. Amendment MW9:1995 was issued to take account of the Construction (Design and Management) Regulations 1994 and Amendment MW10:1996 dealt with more insurance changes; Amendment MW11:1998 made significant changes arising out of the Latham Report and the Housing Grants, Construction and Regeneration Act 1996. In effect, MW 80 in its revised form was a completely new set of contract conditions. The headnote to the form as issued in 1980 set out its purpose:

The Form of Agreement and Conditions is designed for use where minor building works are to be carried out for an agreed lump sum and where an

Architect or Contract Administrator has been appointed on behalf of the Employer. The Form is not for use for works for which bills of quantities have been prepared, or where the Employer wishes to nominate sub-contractors or suppliers, or where the duration is such that full labour and materials fluctuations provisions are required; nor for works of a complex nature or which involve complex services or require more than a short period of time for their execution.

Users found this headnote misleading and it was withdrawn in August 1981 and replaced by Practice Note M2, which is much more indicative of the scope of the contract.

A new form of contract was printed at the end of 1998. It was based on the 1980 edition with JCT Amendments MW1 to MW11, together with some changes and corrections. The new form was still recognisably derived from MW 80 and it was referred to as MW 98. There were a further five amendments to MW 98 since publication. In 2005, the whole suite of JCT contracts was revised and MW 98 became MW with a variant (MWD) incorporating a contractor's designed portion (CDP). The contract was revised in 2007 and again in 2009. In 2011 MW and MWD were revised again and reprinted to take into account the changes made by Part 8 of the Local Democracy, Economic Development and Construction Act 2009. For work carried out in Northern Ireland a short adaptation schedule is available.

1.3   When to use MW and MWD

The criteria for use of the form are set out inside the front cover of the form. They are set out below with comments:

- *Where the work involved is simple in character.* The form is relatively short and it is not sufficiently detailed for use where anything complex, whether in the structure of the building itself or in the services, is envisaged. More complex buildings often raise issues of valuation, extensions of time and financial claims. Even with the terms which the law will imply into this contract (see Chapter 3), it is not suitable for complex work.

- *Where the work is designed by or on behalf of the employer.* In the MWD, unsurprisingly, there is an addition dealing with the situation where the contractor is required to design part of the work. MWD is a much needed alternative. Some architects and quantity surveyors believe that contractor's design can be imported by means of a carefully inserted clause in the specification; that is not correct.

- *Where the employer is to provide drawings and specification or work schedules to define adequately the quantity and quality of the work.* It should be noted that the contractor's obligation is to carry out the work shown collectively in the contract documents. There is no provision for guaranteed quantities as in some of the forms intended for larger Works.

- *Where it is intended that an architect/contract administrator is to administer the conditions.* The phrase 'contract administrator' is sufficiently wide to include any person so designated by the parties. In theory, this could even be the employer, but there are problems with that approach (see Chapter 6).

When not to use

The contract is said not to be suitable if bills of quantities are required, although often the work schedules appear to be bills of quantities by another name. It is not suitable for use if it is desired to have some of the work carried out by named specialists, because there are no clauses to govern the process. Clearly, the use of nominated or named sub-contractors as such would require substantial amendments to the form as printed, although MW envisages that the contractor may sub-let with the architect's consent. It has sometimes been suggested that the control of specialists can be achieved by means of the employer contracting directly with the specialist. This may have other unfortunate consequences. Possible ways of dealing with the situation are explored in Chapter 9.

The contract is not suitable where detailed control procedures are needed, because there are no detailed procedures. Detailed control would be needed for a complex building.

MW is not suitable if the contractor is to design part of the Works; MWD should be used. However, MWD cannot be used as a design and build contract. That is a pity, because the industry is sorely in need of a design and build contract for simple work.

The criteria no longer, as in former editions, give advice about the value of contracts for which MW is suitable. The value was never as important as the simplicity of the work and the contract period. No period is suggested either. Very tentatively, an upper limit of about £150,000 and a contract period no longer than six months might be suggested so far as the current form is concerned.

MW not that simple

MW should not be used merely on account of its apparent simplicity or because, sensibly, the architect dislikes the complex administrative procedures of the Standard Building Contract (SBC) or the more detailed provisions of the Intermediate Building Contract (IC). The brevity and simplicity of MW is more apparent than real because its operation depends to a large extent on the gaps in it being filled by the general law. Some of the more obvious gaps can be plugged by drafting (or getting an expert in construction contracts to draft) suitable clauses. To take one example: nowhere in MW is there any provision dealing with contractor's 'direct loss and/or expense' claims as found in SBC and IC, except for the provisions which require the valuation of variations to include any direct loss and/or expense incurred by the contractor due to regular progress of the Works being affected by compliance with a variation instruction.

3.6.3

This does not mean that the contractor must allow for the possibility of claims in its tender price or that, in appropriate circumstances, it cannot recover them. It merely means that there is no contractual right to reimbursement and that the architect has no power under the contract to deal with them. As explained in Chapter 23, the contractor can pursue such claims in adjudication, arbitration or by means of legal action. In Chapter 15 there are appropriate suggestions for dealing with this familiar construction industry problem.

1.4 How to use

Within its limitations of use (see above in this chapter), MW can be used with a wide variety of supporting documents. Taken together, they are termed the contract documents. In principle, they may consist of any documents agreed between the parties to give legal effect to their intentions.

A number of options are set out in the second recital and they may be conveniently considered as follows:

- The contract drawings and the contractor's schedule of rates;
- The contract drawings and the specifications priced by the contractor;
- The contract drawings and work schedules priced by the contractor;
- The contract drawings, the specification and the schedules – one of which is priced by the contractor.

One of these options, together with the Agreement and Conditions, form the contract documents. They must *each* be signed by or on behalf of the parties.

Before embarking on a project, the options must be studied carefully to arrive at the combination which is most suitable for the work. Note that the third recital provides that the contractor must price either the specification or the work schedules or provide its own schedule of rates.

The contract drawings

On very small Works, for which of course MW is very well suited, it may be quite acceptable to go to tender merely on the basis of MW and drawings. This system can be very satisfactory, provided that all the information required by the contractor for pricing purposes is included on the drawings. The architect is not precluded from issuing further information by way of

2.3 (2.4) clarification as necessary; indeed it is the architect's duty to do so, but the contractor cannot expect the architect to provide every detail no matter how minute. The contractor is expected to use its own practical experience in con-

1 structing the works.

Since there is no specification or schedules, the contractor will be expected to provide its own schedule of rates. If this is not intended, the third recital must be deleted in its entirety. The significance of the priced document is that it is to

3.6 be used to value variations, if relevant.

The contract drawings and the specification priced by the contractor

In practice, this is a very common way of dealing with small projects. If the specification is to be priced, great care must be exercised in preparing it. This system involves the contractor in taking off its own quantities with reference to both drawings and specification and the cost of so doing is likely to be reflected in the tender figure. The organisational expertise which is incorporated into the specification will determine how useful the priced document will be in the valuation of variations.

The contract drawings and work schedules priced by the contractor

This is the variant of the last system. Since the Works must be specified somewhere, it is likely that the specification element will be incorporated into the schedules. Alternatively, depending on the type of work, it may be feasible to put all the specification notes on the drawings. The schedules will normally be quantified, making this system much easier to price from the contractor's point of view. Indeed, often the schedules are really bills of quantities under another name. The contractor needs to take care, however, that its price is inclusive of everything required to carry out the Works. This is true even if some items are missed off the schedules but shown on the contract drawings. The point can be
2.4 (2.5)   a difficult one. It is discussed in section 1.8 below: correction of inconsistencies.

The contract drawings, the specification and the work schedules, one of which is priced by the contractor

In practice, this combination would be used on larger Works when the work schedules would take the form of bills of quantities. The contractor would then normally price the schedules rather than the specification. Once again, the contractor must take care that it prices for everything the contract requires it to do since even if the work schedules are in the form of bills of quantities, the employer does not warrant their accuracy, neither does the contract provide that the quantities, if given, take precedence over drawings or specification. Thus, if five doors are listed in the schedules, but it is clear the drawings show seven doors, the contractor must price for seven doors and will be taken to have done so.

In principle, a work schedule is always to be preferred over a schedule of rates. The former is capable (or should be capable) of being priced out and added together to arrive at the tender figure. A difficulty may arise because it requires a broad measure of agreement on the method of carrying out the Works. The contractor will have difficulty in pricing a schedule of work if it considers that a totally new approach will show greater efficiency. Some two-stage method of tendering will probably yield best results in such cases when the contractor can be expected to input its suggestions before the schedules are drawn up. Whether two-stage tendering is justified on small projects is another matter.

On the other hand, the figures in a schedule of rates cannot be added together to give the tender figure, and the contractor's own schedule cannot be accepted unless it has justified the calculation of the overall sum from the basis of the schedule. To do otherwise would reduce the valuation of variations to a farce. A schedule of rates is most useful where the total content of the work is not precisely known at the outset. MW can be used in this way, with a little adjustment, but it is better to consider some other forms such as SBC With Approximate Quantities or the Prime Cost Contract (PCC).

Second recital   The numbers of the contract drawings must be inserted in the space provided. On large contracts, when bills of quantities are used, it is usual to designate as contract drawings only those small-scale drawings which show the general scope and nature of the work. Under this contract, however, the situation is very different. The total number of drawings is likely to be relatively small and the contractor will need all of them in order to prepare its tender. Since the contractor's basic obligation is to 'carry out and complete the Works in a proper and workmanlike manner and in compliance with the Contract Documents

2.1.1   ...' the architect must be sure that the contract documents taken together do cover the whole of the Works. Therefore, the contract drawings must be:

- The drawings from which the contractor obtained information to submit its tender.
- Sufficiently detailed so that, when taken together with the specification and/or work schedules, they include all workmanship and materials required for the project.

2.3 (2.4)   The further information which the architect must provide may consist of drawings, details and schedules. Provided that they are merely clarifying existing information, there is no financial implication. If, however, they show different or additional or less work or materials than those shown in the contract documents, the contractor will be entitled to a variation on the Contract Sum. None of the 'further information' constitutes a contract document.

1.5   What is the contract?

What MW and MWD say

There are still a great many people who think that when reference is made to the 'contract' the reference is to the printed form MW or MWD properly filled in and signed by the parties. Actually, the contract usually consists of a bundle of documents which are referred to in MW or MWD as the 'Contract

Second recital   Documents'. The contract documents consist of the contract drawings, the
(third recital)   specification, the work schedules, the printed form and, if neither the specification nor the works schedules have been priced, a schedule of rates. If MWD is being used the documents will also include the Employer's Requirements.

Incorporating other documents

In addition, other documents may be made part of the contract by what is termed 'incorporation'. To do that, it is necessary to clearly state in the printed form that these other documents are to be incorporated in the contract. A good example of incorporation of other documents is contract drawings which must *Second recital* be listed leaves space for the contract drawings to be listed but, if there is insuf- *(third recital)* ficient space to list them, they may be listed on a separate sheet which must be clearly identified in the space and attached to the back of the printed form. It is a good idea and avoids any doubt if the sheet is signed and dated by the parties although the form does not mention that.

It is unfortunately common for document-happy employers and their advisors to try to incorporate all kinds of other documents. For example, if there has been correspondence between employer and contractor after tender stage, it might be thought useful to include it. Generally, the inclusion of extraneous pieces of correspondence as part of the contract only serves to confuse the issues if a dispute arises and may even be a cause of the dispute as the parties form radically different ideas of what they said, or intended to say, in the correspondence. If the contents of letters are judged to be so important, it is better to have what was said formally agreed and the contract documents amended accordingly.

Occasionally, an attempt will be made to incorporate the whole of a contract simply by reference to it in a letter, e.g. 'on the JCT Minor Works Contract terms and conditions'. That cannot work properly, because it overlooks the fact that there are many blanks in the printed form (such as the recitals and the contract particulars) which must be filled in before the contract makes any sense.

Incorporating terms by reference is a dangerous practice. It ignores the fact that not only may there be earlier versions still in existence but there may be different editions of a contract so that it is impossible to say with certainty which amendment applies. The result is that since the parties are rarely in agreement regarding which contract is referred to, there is in many cases no contract and the one who has done work will simply be entitled to a *quantum meruit* (see Chapter 3, section 3.5).

1.6   How to complete the contract form

It cannot be stressed too much that the form must be completed carefully. Although it may not very much appeal, it is a job for the architect or other person who will actually administer the contract. It is unusual for a quantity surveyor to be appointed on this contract but where a quantity surveyor is appointed, possibly to produce the work schedules, it is useful to seek his or her advice. However, it is not a task for the quantity surveyor alone, still less for the employer's solicitor who may or may not have a full understanding of MW and MWD but who will certainly not have experience in administering a building contract in progress.

It should be noted that although the specification or work schedules usually state in the preliminaries section how the contract form is to be completed, most of that information must be provided by the architect even if a quantity surveyor produces the schedules. That means that the architect is effectively completing MW or MWD when providing that information. Theoretically, when an acceptable tender is received, the architect simply copies the information in the preliminaries into the contract form. Realistically, the process of accepting a tender may not be straightforward, certain things may be changed and the transfer of information from preliminaries to the contract form may be subject to adjustment. This section explains how to complete the contract form and the actions are summarised in Table 1.1.

First page

Conveniently, the part of the Contract requiring completion is at the beginning of the document and extends to and includes the attestation. Before and while completing the contract, it is essential to read the relevant clauses with great care to see that, in filling in the blanks and deleting, the results are exactly what is intended. The document is complicated and the architect should never 'take a stab' at an entry. If there is any doubt, the advice of a contract expert should be obtained. This may entail rather more than referring to the quantity surveyor (if appointed).

On the first page the date of the agreement should be left blank until the last party signs. The date must be the date on which the last party signs even if that is when the Works are very far advanced or even completed. Obviously, every effort must be made to have the contract signed before work starts on site. It becomes more difficult afterwards.

Recitals

The first recital is important, because it gives a description of the Works and the location. The description must be entered carefully but not in too great detail, because there must be scope for the architect to issue instructions requiring variations without altering the character and scope of the Works.

MWD second recital requires the insertion of the parts of the Works which the contractor is to design. The description must be clear and unambiguous. For example, if the contractor is to design the heating system, it may be enough to simply say 'heating system', but the clearer the description the less chance there is of disputes later. So instead of simply inserting 'heating system', it would be prudent to make clear that it includes all boilers, pipes and radiators including space and domestic hot water heating and anything else particular to the Works. If the space is insufficient, reference must be made to a separate sheet firmly attached to the contract, signed and dated by the parties.

The second recital, MWD third recital, requires the contract drawing numbers to be entered. If there is insufficient space, the reference to a clearly identifiable list of numbers of all contract drawings must be inserted. The list must be firmly attached to the contract, signed and dated by the parties. It is essential that the contract drawings are exactly the same as the tender drawings. Any post tender amendments should be shown by means of an addendum sheet signed, dated and attached to the contract. Depending on what documents are to be attached to the contract, the reference to drawings, specification or work

Table 1.1   Filling in the MW form (MWD variations shown).

Item or clause	Comment
Articles of Agreement	Names and addresses of the parties inserted. Date to be inserted when the last party executes the contract.
First recital	The description of the Works must be brief but sufficient to identify them clearly.
Second recital (Third recital)	The contract drawing numbers must be filled in or an attached list of numbers identified. Delete inappropriate documents.
Second recital (MWD)	Insert extent of CDP work or refer to and identify attached signed sheet.
Third recital (Fourth recital)	Delete inappropriate documents.
Article 2	Contract Sum in words and figures inserted.
Article 3	Insert the name and address of the architect or contract administrator (if not an architect) and delete the alternative.
Article 4	Insert the name of the principal designer if it is not the architect and if all the CDM Regulations apply, otherwise delete.
Article 5	If all the CDM Regulations apply, insert the name of the principal contractor if it is not the main contractor, otherwise delete.
Article 6	This may be deleted if the contract is not a 'construction contract' as referred to in the Housing Grants, Construction and Regeneration Act 1996 (as amended).
Article 8	Amend reference to 'English' to a different jurisdiction if required.
Contract particulars Fourth recital (Fifth recital)	Insert the date chosen as the base date, having considered schedule 2, paragraphs 1.1, 1.2, 1.5, 1.6, 2.1 and 2.2.
Contract particulars Fourth recital (Fifth recital)	State whether Employer is a 'contractor' under the Construction Industry Scheme.
Contract particulars Fifth recital (Sixth recital)	State whether notifiable under the CDM Regulations.
Contract particulars Sixth recital (Seventh recital)	If there is a framework agreement, state date title and parties otherwise delete.
Contract particulars Seventh recital (Eighth recital)	State whether some or all of the supplemental provisions apply and if paragraph 6 applies state the senior representatives of employer and contractor charged with resolving potential disputes.
Contract particulars Article 7	Delete to show if arbitration is to apply. If no deletion, legal proceedings will apply.

Table 1.1   (Continued)

Item or clause	Comment
Contract particulars 2.2 (2.3)	Insert date of commencement. Do *not* insert 'to be agreed'. Insert date for completion. Do *not* insert 'to be agreed'.
Contract particulars 2.8 (2.9)	Insert amount of liquidated damages and the period, e.g. 'day' or 'week'. Do *not* use the words 'per week or part thereof'.
Contract particulars 2.10 (2.11)	The default period for the rectification period is three months. If a different period is required, it should be inserted.
Contract particulars 4.3	The default figure is 95%. If desired, a different figure may be inserted.
Contract particulars 4.5	The default figure is 971/2%. If desired, a different figure may be inserted.
Contract particulars 4.8.1	The default period is three months. If a longer or shorter period is desired, it should be inserted.
Contract particulars 4.11 and schedule 2	Delete if fluctuations are not to apply.
Contract particulars 4.11 and schedule 2, para. 13	Enter the percentage if an additional sum is to be paid, otherwise enter 'Nil'.
Contract particulars 5.3.2	Insert the amount of insurance cover required.
Contract particulars 5.4A, 5.4B and 5.4C	For new Works choose 5.4A. For works to existing structures choose either 5.4B, or 5.4A and 5.4C and delete the other(s).
Contract particulars 5.4A.1, 5.4B.1 and 5.4C	If the percentage to cover professional fees is not to be 15%, insert another figure.
Contract particulars 7.2	Insert name of adjudicator if desired or write 'not used'. Delete four of RIBA, RICS, CC, AICA, CIArb.
Contract particulars Schedule 1, para. 2.1	Delete two of RIBA, RICS, CIArb.
Clause 1.7	Amend if not the law of England.

schedules may be deleted. Every contract document must be signed and dated by both parties to avoid any later dispute regarding what is or what is not a contract document. In practice, this means signing every drawing and the cover of the specification and/or schedules. It is suggested that each document is endorsed: 'This is one of the contract documents referred to in the Agreement dated…' or that some other form of words to the same effect be used.

The third recital (MWD fourth recital) confirms that the contractor has supplied either a priced specification or work schedules or a schedule of rates. The two which do not apply must be deleted. No entries or deletions are required in the fourth to seventh recitals (MWD fifth to eighth recitals) inclusive which refers to matters identified in the contract particulars.

Articles

Article 1 simply sets out the contractor's obligation in brief and requires no insertions. The Contract Sum (excluding VAT) must be inserted in Article 2 in words and figures. The Contract Sum is always this figure. If the contract states that amounts are to be added to, or deducted from it, the figure becomes the adjusted Contract Sum. The name and address of the architect must be inserted in Article 3. That is usually the name of the firm. The alternative description, 'Contract Administrator' is used here and throughout the contract where the person is not entitled to use the description 'Architect' Therefore if the name of an architect is inserted, the words 'Contract Administrator' must be deleted. If the name of someone other than an architect is inserted, the word 'Architect' must be deleted. If the architect ceases to act as contract administrator for any reason, the employer must nominate a replacement within 14 days.

If the CDM Regulations apply, Article 4 states that the architect will act as the principal designer, unless the name of another person is entered as principal designer. If another person is entered, delete the words 'Architect/Contract Administrator'. Article 5 states that the contractor will be the principal contractor for the purposes of the CDM Regulations unless another person's name is inserted. If another person is entered, delete the word 'contractor'.

Article 6 does not require any insertions or deletions. It merely states the right of either party to refer any dispute to adjudication. If the employer is a residential occupier, Article 6 may be deleted, but not otherwise. The adjudicator can be named in the contract particulars, agreed between the parties or nominated by one of the nominating bodies listed. Articles 7 and 8 refer to the right of the parties to refer disputes to arbitration or legal proceedings respectively. The choice is to be made in the contract particulars. Some people delete the Article not required, but that is not necessary. What is important is that the choice is correctly indicated in the contract particulars. If Article 8 is chosen, amend 'English courts' to some other jurisdiction if appropriate

Contract particulars

It is important that the contract particulars are completed so as to correspond precisely with the information given to the contractor at tender stage. If any changes were agreed for example, in the date of possession, the changed information must be faithfully recorded. The words 'to be agreed' or sometimes 'TBA' should be avoided. Inevitably they are never agreed. The best chance of agreement is always before, not after, the contract is signed. In some instances, leaving the entry blank will trigger a default position which may not be what is required. Most of the entries are self-evident, but special care should be taken with the following:

■ The base date must be inserted (see Chapter 2, section 2.7).

■ Whether the employer is a 'contractor' for the purposes of the Construction Industry Scheme. Generally, an employer under MW or MWD will not be a contractor in this sense, but to make sure it is best to check with the legislation.

Seventh recital
(eighth recital)

- Note the default position that if no deletions are made against a paragraph, that paragraph is to apply.

Article 7

- It should be noted that a deletion must be made to show whether arbitration is or is not to apply. If no deletion is made, arbitration will not apply.
- Enter the dates by which the Works should be commenced and completed or a method of unambiguously calculating them, e.g. 'Date for Completion is 15 weeks from the Date for Commencement'. There is no real way of

2.2 (2.3)

avoiding writing the 'Date for Commencement'. TBA or 'two weeks after notice to commence' should be avoided.

- When entering the rate of liquidated damages, the period concerned, e.g. 'per day' or 'per week' must be stated. The phrase 'per week or part thereof' must not be used. It does not, as commonly thought, mean that the damages are to be calculated pro-rata for part of a week. It means that the same damages are applicable for just one day as for a whole week; which is probably not enforceable. If 'pro-rata' is intended write: 'per week or pro-rata'. If the word 'Nil' is entered for the rate, no liquidated damages will be recov-

4
2.8 (2.9)

erable. If the rate is left blank the position is uncertain. It may be that no damages can be recovered.

- Decisions about levels of insurance and the like should be the subject of

5.3, 5.4A. 5.4B,
5.4C, 5.5
5

discussion between the employer and an insurance expert such as an insurance broker. Few architects have any degree of expertise in insurance matters and should refrain from advising the employer.

- An adjudicator can be named in the contract particulars. Whether or not that

7.2

is done, the chosen nominating body must be indicated by deleting the others.
- If arbitration has been chosen instead of legal proceedings, one of the bodies must be chosen to appoint the Arbitrator by deleting the others. The default appointer is the President or a Vice-President of the Royal Institute of British Architects.

Attestation

There are three attestation pages preceded by a page of mercifully brief notes. The attestation pages are simply the pages on which the parties to the contract sign to signify their agreement to the contract terms. 'Attestation' means the witnessing of an act or an event. When the parties attest, they are said to 'execute' the contract. There are three pages, because there are two options and four ways in which to carry out the second option. The first option is to execute the contract under hand (referred to as a 'simple' contract). The second option is to execute the contract as a deed (referred to as a 'specialty' contract. The differences are set out in Chapter 1, Section 1.1). The notes are very clear.

The conditions

It may be appropriate to amend certain of the clauses in this section:

1.1

- Amend the definition of Public Holiday if a different definition is applicable.

1.7

- Amend the clause if the law of England is not required.

1.7 Priority of documents

It is important to understand the relative importance of the various contract documents. This is crucial when there is a dispute and it is necessary to discover what was agreed. Unfortunately, it is common for parts of documents to be in conflict with other documents. That is particularly the case when they have been assembled in a hurry, perhaps after a long and difficult negotiation following tender stage.

1.2

The most important document is the printed form containing the articles, recitals, contract particulars and the conditions. If there is a conflict between this document and any other document, the printed form takes precedence. For example, if the specification states that the employer may terminate the contractor's employment after only five days instead of seven days from the date of the architect's default notice that will not be effective unless the change has actually been made in the clause in the printed form.

1.2
6

It is not permissible to simply pick out a clause in isolation in order to prove a point. Unless they are in conflict, all the parts of the printed form must be read together, because some clauses may qualify others. For example, the contractor's obligation to pay the amount stated as the Contract Sum must be read with the clause which empowers the architect to issue instructions requiring

3.6

variations and then to value them.

Work included

Articles 1
2.1.1 (2.1)

Another related question concerns what work has been included in the Contract Sum. Is it the work shown on the contract drawings or is it the work in the specification or in the schedules of work? The answer is that the contract requires the contractor to carry out and complete the Works in compliance with the contract documents, the construction phase plan, statutory requirements and, under MWD, the Employer's Requirements. Therefore, if something is included in one of the documents but not in the others, the contractor is obliged to include it in the construction and is deemed to have included it in the Contract Sum. This can lead to problems in practice, because contractors usually simply price the specification or the schedule of works. Many contractors do not realise that even if these documents do not include something present on a drawing, they must price for it. So if the contract documents comprise drawings and specification and the architect has shown a wood block floor finish on the drawings but missed it from the specification and the contractor has only priced what is in the specification, the contractor will be obliged to provide the wood block finish as specified at no additional cost.

1.8 Inconsistencies and divergences

Errors and discrepancies in the Employer's Requirements

2.1.2
7

Under MWD, the contractor is not responsible for anything in the Employer's Requirements. In particular, it is not responsible for checking to see that any design contained in the Employer's Requirements is 'adequate'. Very often

Employer's Requirements include very substantial design of the project leaving the contractor with little more than to detail elements of construction. What this means is that the contractor is entitled to assume that the design will work and to proceed accordingly.

It is doubtful that the contractor is entitled to ignore a design error that is, or becomes, obvious. Indeed, it may well be that the contractor has already dealt with some design flaws in the Employer's Requirements when preparing the tender. However, if any inadequacy in the Employer's Requirements comes to light, the correction of a design error on the part of the employer could well involve quite significant variation and cost.

Discrepancies in general

If there is any inconsistency between the contract drawings, the specification, schedules of work and Employer's Requirements (if under MWD), it must be corrected, presumably by the architect and the change must be treated as a variation. Under MWD, if there is an inconsistency between any of the documents prepared by the contractor for the CDP work, the contractor must correct it at its own cost after the architect has first expressed satisfaction with the contractor's proposal. The architect' satisfaction must be reasonable (see Chapter 5). There is nothing in the contract which compels either architect or contractor to notify the other about inconsistencies, but no doubt the contractor will be alive to inconsistencies in the architect's documents and the architect will keep a close eye on the CDP documents. However, if neither architect nor contractor notice an inconsistency in the architect's documents until the problem has taken on three-dimensional form in the building itself, the employer will normally be responsible for the cost of rectification. If neither architect nor contractor spot an inconsistency in the contractor's CDP documents, the contractor will have to rectify the problem at its own cost.

Importance and priority

The contract documents provide the only legal evidence of what the parties intended to be the contract between them. They are, therefore, of vital importance. In the case of dispute, the adjudicator, the arbitrator or the court will look at the documents in order to discover what was agreed.

The contractor is obliged to carry out the Works in accordance with the contract documents. The question often arises: what is the position if the documents are in conflict? The printed form must be read as a whole; nothing contained in the contract drawings, the specification, the schedules, any framework agreement or, if MWD is being used, the Employer's Requirements will override or modify the printed conditions. If, therefore, the specification was to contain a clause purporting to remove the contractor's entitlement to extension of time due to late receipt of information, it would be ineffective because of this provision, which has the effect of reversing the normal rule that type prevails over print. The effectiveness of a clause worded in this way has been upheld in the courts on many occasions.

Although the contract effectively sorts out priorities as between the printed form and the other contract documents, it gives no further guidance as far as priorities among the other contract documents are concerned. Clause 2.4 simply states that Any inconsistency in or between the contract drawings and the contract specification and the work schedules must be corrected and if such correction results in addition, omission or other change, it must be treated as a variation.

2.4 (2.5)

The particular circumstances of each case will determine exactly how the inconsistency is to be treated. Two main types of inconsistency are common:

1. Where workmanship or materials are covered in one of the contract documents, but omitted from the other
2. Where workmanship or materials shown in one of the contract documents are in conflict with what is shown in the other.

This consideration will be confined to a contract based on drawings and specification which the contractor has priced. If a schedule is also included, the principle is the same but the facts may be more complicated. The first thing to establish is what the contractor has legally agreed to do. It is this:

The Contractor shall carry out and complete the Works in accordance with the Contract Documents.'

Article 1

2.1.1 (2.1)
Second recital
(third recital)

The same obligation to comply with the 'Contract Documents' is repeated and that clearly means the documents noted in the contract, that is, in this case, the contract drawings, the contract specification and the conditions. If, therefore, the inconsistency is, for example, the fact that a handrail is shown without bracket on the drawings, but brackets are specified in the specification, or vice versa, it will be deemed that the contractor has allowed for the brackets in its price. This is the case even if the brackets are not in the specification, which the contractor has priced, but are shown on the drawing. In this example, even if the brackets are not shown or mentioned on either document, it is likely that the contractor must supply brackets (presumably the cheapest it can find to do the job) at no additional cost, because it will usually be implied that the contractor has allowed in its price for everything which are understood must be done. Much, however, will depend on any general terms which have been included in the specification. An expression such as 'The whole of the materials whether specifically mentioned or otherwise necessary to complete the Works must be provided by the contractor' would tend to place the responsibility for omissions squarely on the contractor provided that they could be considered 'necessary' to complete the Works. In this case, brackets are obviously necessary to support the handrail.

10

If the documents are in conflict, the position is rather complicated. Since neither drawings nor specification have priority, it is not clear as to which of them the contractor has had reference in formulating the price. It is tempting to consider that the key document is the specification, since the contractor has priced it. This is a wrong view of the situation. In pricing the specification, the contractor must consider all the documents. It is thought that, if the documents are in conflict, it is for the architect to instruct the contractor as to which document

Article 1 and
clause 2.1.1
(2.1)

is to be followed in the particular instance, and the contractor is not to be allowed any addition to the Contract Sum nor is there to be any omission, the contractor being deemed to have included for whichever option the architect chooses.

If, however, the architect solves the problem by omitting the work or changing it to something other than is contained in either of the contract documents, *3.6*   it must be valued in accordance in the usual way. Although this view may be thought to impose undue hardship on the contractor in certain circumstances, particularly as the inconsistency is due to the architect's oversight, it is the only interpretation that takes account of the contract as a whole. The contractor bears a great responsibility to examine all the documents thoroughly when pricing. This analysis has an odd result. Although the contract refers to treatment *2.4 (2.5)*   of the correction as a variation, it is difficult to envisage any situation in which the contractor would be entitled to have such a variation valued at anything other than a £nil amount. If the contract is on the basis of drawings and priced schedules which are in fact fully developed bills of quantities, the situation remains the same. That is because under the MW and MWD contracts the employer does not warrant that the bills of quantities are accurate. This, of course, is in complete contrast to SBC and points out one of the great dangers to the contractor in using this form for Works of a greater value than that for which it is intended.

1.9 Custody and copies

There is no specific provision in the contract regarding the custody of the contract documents and the issue of copies to the contractor. It is probably sensible for the architect to keep the original, but if the employer wishes to have it, the architect should be sure to have an exact copy. Although there is no express requirement in the contract, it is good practice for the architect to make a copy for the contractor and certify that it is a true copy. This can be simply done by binding all the contract documents together and inscribing the certificate on each document including the drawings. It is sufficient to state 'I certify that this is a true copy of the contract document dated …'.

It must be implied in the contract that the architect provides the contractor with two copies of the contract drawings and specification and/or schedules, otherwise the contractor would be unable to carry out the Works. It is usual to provide such copies free of charge.

Any further drawings, details or schedules which the architect is to provide *2.3 (2.4)*   are not contract documents. They are intended merely to amplify or clarify the information in the contract documents. The architect is obliged only to supply such drawings as are 'necessary for the proper carrying out of the Works'. It is 11   an implied term of the contract that such additional information will be **issued** 12   at the correct time and that the information will be correct. Failure to do so is a 13   breach on the architect's part for which the employer may be responsible. The contractor would have a legitimate claim for an extension of time without the necessity for any prior application for the information. If the contractor suffers disruption, it may also have a claim for damages for breach of contract which it could pursue at common law.

1.10   Limits to use

The contract contains no specific terms to safeguard the architect's interests in the drawings and specification and there is no express prohibition on the employer from using the contractor's rates and prices for purposes other than this contract.

The general law, however, covers the position. The architect retains the copyright in drawings and specification (unless specifically relinquished) and neither the contractor nor the employer may make use of them except for the purpose of the project. Strictly, the architect may ask the contractor to return all copies after the issue of the final certificate, but in practice it is seldom worth the trouble to receive a collection of torn, stained and unreadable pieces of paper.

The confidentiality of the contractor's rates and prices is safeguarded by the general rule that neither employer nor architect may divulge confidential information to third parties. This is especially true when two parties are bound together in a contractual relationship and to divulge the information would clearly cause harm. The contractor's prices are a measure of its ability to tender competitively and secure work. To divulge its rates to a competitor is a serious matter. In practice, it is not easy for the contractor to ensure, for example, that the quantity surveyor, if appointed, does not make use of its prices to assist in estimating the cost of other contracts, but that is probably of little consequence.

It may be thought prudent to include a clause in the specification to cover the limitations on the use of documents. It does no harm to remind the parties of their obligations in this respect. Clause 2.8.3 of IC 11 with appropriate adjustments could be used.

1.11   Notices, time and the law

All communications referred to in the contract must be in writing. That includes all notices, certificates and instructions. In the case of instructions it should be noted that the contract does allow them to be issued orally, but they must be confirmed in writing within two days. If the method of service of a notice or any other document is not specified in the contract, it may be served by any effective means. That is by any means which has the desired effect. It can be served to any agreed address. If the parties, for one reason or another, cannot agree the address in each case, service can be achieved by pre-paid post to the intended recipient's last known main business address. If the recipient is a corporate body, the address should be its registered or principal office. There appears to be nothing to prevent second class postage being employed although it would not be in the sender's interest to do so. The Civil

Procedure Rules are only binding in legal proceedings, but they provide a useful set of guidelines. Under the CPR:

First class post:	is deemed served the second day after it was posted.
Delivery by hand:	is deemed served the day after it was delivered or left at the permitted address.
Fax:	is deemed served, if transmitted on a business day before 4pm, on that day; or in any other case, on the business day after the date on which it is transmitted.
Other electronic method:	is deemed served on the second day after the day on which it is transmitted.

If a document is served personally after 5 pm on a business day or at any time on a Saturday, Sunday or a bank holiday, it will be treated as being served on the next business day ('business day' being any day except Saturday, Sunday or a bank holiday).

Where anything is stated to be done within a number of days after or from a certain date, the counting of days begins immediately after that date. Therefore, 'seven days after receipt of a notice' means that if the notice is received on the 4th of the month the seven days start on the 5th and expire at the end of the 11th of that month. Any day which is a public holiday is excluded from the reckoning. Therefore, in the example, if the 8th was a public holiday, the seven days would expire at the end of the 12th of the month.

1.4
15

1.7

The law applicable to the contract is the law of England. That is the case whatever may be the nationality of any of the parties to the contract or anyone connected to it. Even if the Works are carried out in Germany under this contract (unlikely) or in France (even more unlikely) the applicable law will still be the law of England. Obviously, the parties may change the applicable law, for example to the law of Northern Ireland.

16

1.5

In common with other forms of contract, MW and MWD exclude the rights of third parties. This effectively reinstates the position as it was before the Act came into force, i.e. only parties to the contract can have any rights under it (see Chapter 2, section 2.6).

1.12   Common problems

Problems can arise with the contract documents which have been prepared for signature or execution as a deed by the architect. The contractor must check them over carefully, checking the printed contract against the information given in the tender documents and noting any discrepancies.

Mistakes in the documents

If the contractor discovers mistakes or inconsistencies, it should not execute the documents until the matter is rectified. Box 1.1 is a suitable pro forma letter for the contractor to send to the architect.

> **Box 1.1   Letter from contractor to architect if mistakes in contract documents and no previous acceptance of tender.**
>
> Dear Sir
>
> PROJECT TITLE
>
> We are in receipt of your letter of the [*insert date*] with which you enclosed the contract documents for us to sign/execute as a deed [*delete as appropriate*].
>
> There is an error on [*describe nature of error and page number of document*]. This is not consistent with the tender documents on which our tender is based and we are not prepared to enter into a contract on the basis of the Contract Documents in their present form.
>
> Therefore, we return the documents herewith and we look forward to receiving the corrected documents as soon as possible.
>
> Yours faithfully

Starting before the contract is signed

Another problem arises where, as is not uncommon, work starts on site before the contractual formalities are completed. It is bad practice to allow this to happen.

It may be that there is already a binding contract in existence, the parties having agreed on the minimum essential terms and there being an unequivocal acceptance by the employer of the contractor's tender. Alternatively, there may be no contract at all, and in that event and if no contract came into being, the work done would have to be the subject of a *quantum meruit* ('as much as it is worth') claim and the contractor would be entitled to a fair commercial rate. However, calculating the amount due is often quite complicated. In general, if a formal contract is signed after work has begun, its terms would be retrospective. That is to say that all the Works, including the work done before the contract came into existence, would be treated as having been done under the terms of the contract, even though some work was carried out before the contract was signed.

It is better to avoid the potential difficulties, and a contractor who is asked to start work before the contractual formalities are completed is well advised to write to the architect appropriately: see Box 1.2.

Box 1.2   Letter from contractor to architect if contractor asked to commence before contract documents signed.

Dear Sir

PROJECT TITLE

Thank you for your letter of the [*insert date*] from which we note that the employer requests us to commence work on site pending completion of the contract documents.

It is our understanding that we are already in a binding contract with the employer on the basis of our tender and the employer's acceptance of the [*insert date*] on terms incorporated by the tender and letter of acceptance.

If the employer will send us written confirmation of agreement with our understanding of the situation as expressed in this letter, we will be happy to commence as requested.

Yours faithfully

2   Some basics

2.1 Works

MW and MWD use the words 'Works' with a capital 'W' when it means the total of all the work, goods and materials which the contractor agrees to provide.

2.2 Drawings

20

It is important to understand what the contract means when it refers to drawings. In this context, a drawing is a visual representation of a building or a part of a building to a particular scale. The contract also refers to 'details' which are usually taken to mean a large scale drawing of a relatively small part of the building. Where 'details' plural is mentioned what it usually intended is an in-depth description rather than a drawing. For example details of sanitary fittings may best be conveyed by means of a schedule and specification.

Types of drawings

So far as MW and MWD are concerned, there are two specific kinds of drawings:

Second recital
(third recital)

(2.1.5)

- Drawings which are part of the contract documents, such as the contract drawings or, under MWD, the drawings produced by the contractor.

- Drawings which are provided by the architect during the progress of the work as part of further information and instructions which the contractor requires to carry out the Works or as part of instructions requiring variations.

2.3 (2.4)

3.6.1

2.3 Copyright

Copyright in the architect's drawings (and in specifications, schedules and the like) is owned by the architect. What that means is that although someone else may actually have, or even own, the drawings, they cannot copy or reproduce what is on them without the architect's permission. The law concerning copyright is quite complex and it is a mixture of legislation and decisions of the courts. If the architect and client have entered into terms of engagement as they should, the terms will usually state that the client has a licence to use the drawings when the architect's fees are paid. A licence is simply permission or authority to do something. If they have not entered into a proper formal agreement, the law will usually imply a term (see Chapter 3) that the client has the right to reproduce the architect's drawings in the form of a building when the client has paid a reasonable fee. Like many terms which are implied, it is not always easy to say what a reasonable fee might be.

The contract says nothing about copyright in drawings and other documents supplied by the contractor. Therefore, the position is left to be covered by legislation. Drawings prepared by the Contractor will remain in its ownership (the usual term is 'vested' to mean that the contractor owns the right). But the Employer, by paying the Contract Sum to the Contractor will have an implied licence to copy and use the documents for any purpose including the main purpose of reproducing them in relation to the Works.

2.4 Specification

A specification is a document which describes in detail, together with the drawings, how a building is to be constructed: quality of workmanship and materials, the way in which the elements fit together and where they occur throughout the building. It may or may not contain quantities.

2.5 Schedules

Architects commonly make use of schedules to a greater or less extent as a useful means of describing the work and materials in a building. Common schedules are ironmongery, sanitary fittings, doors, windows and colour schedules. Obviously, almost anything can be scheduled. A work schedule may be part of the contract documents. Work schedules are usually lists of all the work and materials needed to construct the Works. The list may use very broad descriptions or it may be very precisely detailed.

It is probably sensible to have a clause in the specification or work schedules to the following effect:

Where and to the extent that materials, goods and workmanship are not fully specified they are to be suitable for the purpose of the Works stated in or reasonably to be inferred from the contract documents and they are to be in accordance with good building practice, including the relevant provisions of current British Standard documents'.

2.6   Privity of contract and third party rights

Privity of contract is the name for an old principle that only the parties to a contract can exercise rights under that contract. For example, a contract between A and B might say that A will do work for B and that B will pay A £500. It might also say that in addition B will pay C £100. If A did the work, A could demand the £500 from B, but C until comparatively recently could not legally demand the £100 because C was not a party to the contract. Conversely, a person who is not a party to a contract cannot have obligations imposed by the contract. Therefore, in the previous example, if the contract had said that C must pay A £100, C could not be made to do so even though C knew that the term had been included. All that appears to be eminently sensible and in accordance with common sense. However, it was thought that a strict application of the privity principle could lead to injustice in certain cases and Parliament decided to change the law. It did this by introducing an Act of Parliament which made it possible for someone who was named in the contract, but not a party to it, to claim under the contract. Therefore, if a contract between A and B stated that A and B would each give C £50, C may, depending on a few other things, be able to claim the money. Most contracts now exclude the Act as do these contracts. Therefore, third parties have no rights under MW or MWD. Third parties can be given rights under some other JCT contracts.

2.7   Base date

This is a date from which certain things are to be measured. In contracts issued before 1987, this date used to be referred to as the 'date of tender'. But there was a problem in setting the date. Very often, the date stipulated for the return of tenders (i.e. the date of tender) had to be extended due to a difficulty with the tendering process. In such cases, the date of tender as stated in what was then known as the 'Appendix' and now as the 'Contract Particulars' was very often a date earlier than the actual date of tender. Therefore, it was wrong to call it the 'date of tender'.

The current expression 'Base Date' much more accurately expresses what was intended. For example, the base date is mentioned in Schedule 2 dealing with fluctuations.

2.8   Common problems

Amendments to the standard forms of contract

It is common practice for standard contracts to be amended before use. The practice is more usual when the SBC, IC and ICD are concerned, but MW and MWD are also amended fairly frequently.

The eminent construction law barrister, the late Donald Keating, said that parties intending to sign up to a construction contract should use either an unamended standard form or their own home-made contract conditions and to attempt a mixture of both was usually a recipe for disaster. In other words, trying to amend a standard form was full of pitfalls. That is very true. I once had a design and build contractor as a client. He was in the habit of offering the JCT Design and Build Contract to his customers. One of the customers engaged a solicitor to vet the contract. The solicitor proceeded to delete everything in the contract referring to contractor design. The project then proceeded with the contractor both designing and constructing but without any contractual liability for his design work. Fortunately, the project went smoothly. On another occasion, an architect carried out some home-made amendments to include bills of quantities in a Minor Works Building Contract. The result was a disaster in every way.

One of the most frequent problems is that, if an amateur tries to amend a standard contract, it is very easy to overlook the effect of the amendment on other clauses. Amending standard contracts is ill-advised, but if it must be done, it should be done only by someone with the appropriate specialist knowledge.

The sub-contractor has a design responsibility but the contractor has no design responsibility

It is amazing how often this problem crops up. Take the case of a contract under MW. The contractor has no responsibility for any design. However, for reasons best known to the employer, an item has been inserted in the specification stating that the heating system has to be designed and installed by a named sub-contractor. The contractor appoints the sub-contractor and the heating system is installed. However, at some stage it is clear that the system does not work properly. After several visits to deal with the problem, the sub-contractor has rather lost interest. On the advice of the architect, the employer engages a mechanical and electrical consultant to investigate and the consultant reports that there is nothing wrong with the installation, but that the design of the system is badly flawed and it will never work correctly. The whole system must be redesigned and almost wholly replaced.

What can the employer do about it? The employer has no contract with the sub-contractor and the contractor has no design responsibility. The employer cannot use one of the MW dispute resolution procedures against the sub-contractor, because the sub-contractor is not a party to the MW. The employer cannot successfully go to adjudication against the contractor because the contractor has no design responsibility and the materials and workmanship are not defective. The sub-contractor's duty to design is owed to the contractor not to the employer. The employer relies upon the goodwill of the contractor to chase the sub-contractor. If the goodwill is not there, the employer is left without any obvious remedy. The rights of third parties are probably excluded from the sub-contract and, even if they were not excluded, they would be unlikely to solve this problem. The employer might try to sue the sub-contractor in tort, but that would be unlikely to succeed unless the employer could show that the sub-contractor owed the employer a duty of care, which is again unlikely.

The problem here is that there is what is commonly referred to as a contractual chain, in other words, linked contracts. In that kind of situation, it is imperative that the rights and obligations of the contractor are stepped down to the sub-contractor and vice versa. In this case, the sub-contractor is responsible for design to the contractor, but the contractor does not have the same responsibility to the employer. Clearly, the answer would have been to have the contract under MWD and make the contractor responsible for the design of the heating system. The contractor should then have been allowed to choose its own sub-contractor to do the design and carry out the work. Then, if there were any problems whether of design or workmanship, the contractor would be responsible.

3 Things you must know

3.1 The Housing Grants, Construction and Regeneration Act 1996 (as amended)

Background

It is surprising how many members of the construction industry have no knowledge at all about the provisions of this legislation. The Act was amended by Part 8 of the Local Democracy, Economic Development and Construction Act 2009 which came into force on 1 October 2011.

Part II of the Act as amended is the important part so far as construction contracts are concerned. What follows is a general survey of the principal provisions of Part II. Part II deals with construction contracts and every architect and contractor should have a copy. It is only a few pages long. Included in the definition of construction contracts is an agreement 'to do architectural, design, or surveying work, or…to provide advice on building, engineering, interior or exterior decoration or on the laying out of landscape in relation to construction operations'.

'Construction operations' are defined in detail. Very broadly they are the construction, alteration, repair, etc. of buildings, structures, roadworks, docks and harbours, powers lines, sewers and the like. They also include installation of fittings such as heating, electrical or air conditioning, external or internal cleaning carried out as part of construction and site clearance, tunnelling, foundations and other preparatory work and painting or decorating. Excluded are such things as drilling for natural gas, mineral extraction, manufacture of certain components, construction or demolition of plant where the primary activity is nuclear processing, effluent treatment or chemicals, construction of artistic works, sign writing and other peripheral installations. More importantly, although the Act has teeth, it does not bite where one of the parties intends to take residence in the subject of the construction operations.

The provisions of the Act used to apply only to 'agreements in writing,' but now oral contracts are now caught by the Act. The Act requires that all construction contracts must include certain provisions. They are:

- *Adjudication*
  Either party must have the right to refer disputes to adjudication with the object of obtaining a decision within 28 days of referral. 'Referral' is both the name of the claim document and the name used for giving the claim to the adjudicator and the other party. A party may give notice of intention to refer at any time and the referral must take place within seven days. The 28-day deadline may be extended by up to 14 days if the referring party wishes or for an indefinite period if both parties agree. The adjudicator may take the initiative in finding out the facts and the law. In other words, the adjudicator does not have to wait until one party raises a point, but can ask for evidence. The adjudicator's decision is binding until the dispute is decided by litigation, arbitration or by agreement. The parties may agree to accept the adjudicator's decision as final. In practice an adjudicator's decision is rarely challenged in the courts. The adjudicator is not to be liable for acts or omissions unless there has been bad faith. The adjudicator has power to correct a decision to remove clerical and typing errors. Provisions dealing with the allocation of costs between the parties must be made after the notice of adjudication has been given. These provisions must be inserted in the contract in writing. Therefore, even if the contract is otherwise oral, this part must be written.

- *Stage payments*
  A party is entitled to stage payments unless the duration of the project is less than 45 days. The parties are free to agree the intervals between payments and the amounts of such payments.

- *Date for payment*
  Every contract must contain the means of working out the amount due and the date on which it is due and must provide a final date for payment. This must not depend on the contractor's performance on another contract. There is a strict notice regime to establish the amount which is to be paid.

- *Set-off*
  Payment may not be withheld, nor money set-off unless notice has been given to pay less than the notified sum particularising the amount to be paid and the basis of calculation. The notice must be given no later than the prescribed period before the final date for payment.

- *Suspension of performance of obligations*
  If the amount properly due has not been paid by the final date for payment and no effective notice withholding payment has been given, a party has the right, after giving a 7-day written notice, to suspend performance of obligations under the contract until payment has been made.

- *Pay when paid*
  Except in cases of insolvency, a clause making payment dependent upon receipt of money from a third party is void. This is intended to outlaw the so-called 'pay-when-paid clause,' but it may not be sufficient to do so. It does not take effect if the third party is insolvent.

If a construction contract does not include these provisions, the Scheme for Construction Contracts (England and Wales) 1998 as amended by the Scheme for Construction Contracts (England and Wales) Regulations 1998 (Amendment) (England) Regulations 2011 comes into effect for the missing provisions just as if the clauses contained in the Scheme were written into the contract. Most standard form construction contracts and all the RIBA terms of engagement comply with the Act and, therefore, the Scheme is not relevant where such terms are used.

3.2   Express and implied terms

It is important to understand the difference between express and implied terms. An express term is whatever is printed or written in the contract or it is a term which the parties have agreed orally at the time they made the contract. Express terms are good, because both parties know what they are and understand what rights or duties they impose. Before considering in detail the express contractual obligations in MW and MWD, it is necessary to consider whether there are any terms which the law will write into the contract.

Implied terms

An implied term is a term of a contract which:

- the parties to that contract did not expressly agree either in writing or orally, and
- is not inconsistent with the other express terms, and
- the law says is part of the contract and is binding on the parties as if it were expressly written into the contract.

Terms may be implied in various ways:

- By Statute, e.g. under the Supply of Goods and Services Act 1982, the Sale of Goods Act 1979 and of course under the Housing Grants, Construction and Regeneration Act 1996 (as amended). Important terms about fitness for purpose, quality, price and the like may be implied into a contract.
- At common law. For example, unless there is any exclusion or express term to the contrary, certain warranties will be implied warranties: that a contractor will supply good and proper materials and will provide completed work which is constructed in a good and workmanlike manner and, if there is no independent designer, that it will be suitable for its purpose to the extent that its purpose has been made known to the contractor.
- A term will be implied where necessary to make the contract work.
- If the parties obviously intended something to be included, but it was not made an express term, then a term to that obvious effect may be implied. In deciding whether or not such a term should be implied the question for the court will be whether it can be said confidently that if at the time the contract was being negotiated someone had said to the parties, 'What will

happen in such and such a case?' They would have replied: 'Of course, so and so will happen; we did not trouble to say that; it is too clear'.

- Unless the express terms of the contract stipulate something different, a term will be implied if the courts have already laid down that in particular types of agreement certain terms will automatically be implied. Therefore, under these contracts there will be an implied term that the employer, and the architect on the employer's behalf, will do everything necessary to enable the contractor to carry out the work and that the Architect will provide the contractor with accurate drawings and information.

- By custom and usage, for example if it has invariably been the longstanding practice in a particular trade, profession or business, then unless the parties have expressly stated to the contrary they will be presumed to have contracted with the intention of operating the agreement according to that custom.

- If the parties have consistently and on a significant number of occasions entered into previous agreements of the same kind on a certain set of terms and conditions, it may be assumed that in future dealings of the same kind, unless expressly provided to the contrary, they are conducting their business on similar terms to those used in their previous course of dealings even though they omitted to enter into a formal contract on that occasion.

There are limits to when terms will be implied. For example, a term will not be implied merely because a court thinks it would be reasonable to do so. Terms of all kinds will generally be implied only if it is not in conflict with or inconsistent with an express term and any implied term must be based on what is presumed to be the intention of the parties.

Apart from legislation, the courts have progressively implied in all building contracts terms that 'the builder will do his work in a good and workmanlike manner; that he will supply good and proper materials; and that it will be reasonably fit for the purpose required".

3.3   Limitation periods

The law recognises that there has to be a limit on the length of time after the event that someone can bring an action through the courts for redress. It would be completely unreasonable and unfair if someone could claim against someone else for something that happened twenty years before. Memories would have faded, documents may have been lost or destroyed and some key witnesses may have vanished or be dead. The Limitation Act 1980 specifies a limitation of six years for actions based on simple contract. What that means is that if one party commits a breach of contract, the other party has six years from the date of the breach in which to issue a claim form (what used to be known as a 'writ').

The effect of the Limitation Act

The Limitation Act has nothing to do with the right to sue. It merely limits the period within which a claimant must start the legal action, otherwise the claimant may be 'time barred' or 'statute barred'. In other words, stopped due to passage of time. Therefore, the claimant can sue even thirty years after the breach of contract, but if the defendant draws the attention of the court to the limitation point, that will usually be end of the proceedings. If the point is not raised, the claimant can carry **on** with the legal action.

In the case of a breach of contract the right to sue begins when the breach of contract takes place, even if the claimant knows nothing about it at the time. Under a building contract, a contractor has two separate obligations. The first is to carry out the work in accordance with the contract during the course of the contract. If the contractor is in breach of this obligation it means that the employer can sue when the breach occurs. The second obligation is to complete the Works in accordance with the contract. A distinct breach of contract does not occur in that respect until practical completion has been certified. Therefore, in the case of a long contract where the breach of contract occurs in the early stages, it is not fatal if no action is taken in the first instance. The limitation period will not start under the second obligation until practical completion has occurred.

How long?

If a contract is entered into as a deed it has a twelve-year limitation period. A simple contract which has only been signed by the parties has a six-year limitation period. Therefore, generally speaking a contractor can be sure that six years after it has completed any particular work no legal action can be successfully brought against the contractor for breach of contract unless the contract is executed as a deed when the period will be twelve years.

It is clear, therefore, that there are advantages for every employer to require the main contractor to enter into a contract as a deed. If the contractor does so, it should ensure that every sub-contractor, whether nominated or domestic, does the same. Of course a sensible contractor will add something to its price for constructing the project if the employer requires a deed.

Exceptions

The position set out above may not apply in two specific instances:

- The limitation period will not start to run if the breach of contract has been concealed by the fraud of the defendant. Normally, in law 'fraud' involves a deliberate intention to cheat. But the courts have tended to interpret this stipulation quite broadly so as to include cases where they believed it would have been unfair to allow the defendant to rely on the statutory defence provided by Parliament. But if a contractor has committed a breach of contract without knowing, it may still be able to rely on the limitation defence.
- The limitation period may be extended to take account of any indemnities which the contractor may have given to the employer under the insurance clauses of the contract.

3.4   Letters of intent

If there is no signed contract

It is not always clear why, but many projects commence on site before the building contract has been signed (strictly 'executed'). It is always dangerous to start work on site before the contract is signed, but to do so before everything is agreed is suicidal for both parties. There is seldom any excuse for failure to execute the contract before commencement. It is one of the most important things each party must do. An employer who, in his or her business life would never dream of putting expensive operations in motion before all the legalities have been checked and double checked, are suddenly consumed with a desire to throw caution to the winds and get on with the exciting process of construction. If a signed contract is not ready to be executed, the employer will often arrange for the contractor to receive what is known as a 'letter of intent'.

What is a letter of intent?

The idea of a letter of intent in its simplest form is that the employer is writing to the contractor to say that the tender (perhaps after negotiation) seems satisfactory and that the employer intends to accept it, but that for one reason or another it cannot be accepted yet and that if the contractor starts work, the employer will see that it is paid at the rates in the bills of quantities until the contract is signed. It may then go on to say that the amount of work the contractor can carry out in this way is limited to £x and that the employer can stop the work at any time. Crucially, the letter should make clear that it is not to be considered as accepting the contractor's tender nor is it forming a binding building contract.

Bearing in mind the dangers, it is surprising how many times contractors start work on the basis of a 'letter of intent'.

Dangers

The dangers are:

- It may be a straightforward simple letter of intent. If so either the employer or the contractor will be able to simply walk away from the work if they are not happy and that could be very awkward for the other. Neither would be in breach of contract if the letter does not bind them to continue working.
- The letter may be badly written so that instead of being a letter of intent, it may be an acceptance of the contractor's tender and a binding contract is formed before the employer, and possibly the contractor, is ready.
- It may be a letter of intent, but may itself contain complicated provisions which amount to the parties entering into a contract under the letter of intent which is disadvantageous to one or the other. Some letters of intent

are many pages long – which misses the point of a letter of intent which should be a short and simple document.

Many cases have been through the courts as a result of contracts being started by letters of intent. Even if a dispute does not get as far as the courts, arbitration or adjudication, it is still something to be avoided. Disputes are common if one or other or both parties do not entirely know where they stand.

The letter of intent may not be binding

A letter of intent may be a continuing offer: 'if you start this work, we will pay you appropriate remuneration'. This creates no obligation on the other party to do the work, but if it does it, there are no express or implied warranties as to its quality. Damages can never be awarded for failure to carry out a unilateral contract, because there is no obligation on the person to whom the promise is made.

3.5   Quantum meruit

It is a common expression in the construction industry to say that someone will be paid a *quantum meruit*. Those words literally mean 'as much as is deserved'. They are often used instead of *quantum valebat* which mean 'as much as something is worth'. It is rare for a distinction to be drawn between the expressions. Most people equate the expression with 'a reasonable sum'. Someone may be paid a *quantum meruit* in four different situations:

- If there is an agreement simply to pay a reasonable sum.
- If work is carried out under what was both parties thought was a valid contract, but which turns out to be void and, therefore, inapplicable.
- If work is carried out by one party at the request of another, e.g. following a letter of intent.
- If work is done under a contract which has no terms stating the amount to be paid. That would never be the case under a standard form contract such as MW or MWD provided all the necessary parts have been completed. It sometimes occurs if a contract has been formed just by exchange of correspondence.

A court will not order a *quantum meruit* to be paid if the contract clearly provides for payments of a specified amount or where, as in JCT contracts, there is a mechanism in the contract to determine the amount. An architect cannot certify anything other than what the JCT contract authorises. Therefore, unless the employer specifically authorises it and the contractor agrees, the architect cannot certify *quantum meruit* payments. The payment on a *quantum meruit* basis does not mean that payment will be calculated on a 'cost plus' basis with an allowance for profit, but instead on the basis of a fair commercial rate.

3.6   Common problems

If the employer and contractor enter into a contract on the basis of a price given by the contractor based on some drawings, will the law imply that the relevant JCT contract is part of the agreement?

That is an interesting idea, but sadly incorrect. The law says that the two parties can, broadly speaking, enter into what kind of agreement they choose and the courts will not intervene to make a new or different contract for them. If the employer realises, too late, that no completion date has been agreed, the law will imply that the contractor must complete within a reasonable time, but no more than that and the other fairly general implications noted earlier (section 3.2 above). Moreover, if the parties have contracted on the basis of the MW contract, it does not matter what some additional clauses in, say, the SBC say. Therefore, a contractor under MW cannot claim loss and expense by referring to the loss and/or expense clause in IC, because it is a different contract which does not apply.

The adjudication is finished and the adjudicator has obviously made the wrong decision. Can the employer or contractor appeal?

There is no appeal against the adjudicator's decision unless it can be shown, for example, that:

- The adjudicator was nominated by the wrong nominating body.
- The adjudicator did not give one of the parties a proper chance to make their case.
- The adjudicator made the decision on the basis of something the adjudicator thought of and neither party had raised the point nor had the chance to comment on it.
- If there was no dispute.
- If there was no construction contract.

However, if the adjudicator is proper appointed, answers the question put in the notice of adjudication and gives both parties an opportunity to have their say, it does not matter if the decision is quite clearly wrong; it will be enforced by the courts. It can be seen that the acceptable reasons for challenging the decision do not concern the decision itself, but what might be termed peripheral matters.

4   Architect's powers and duties

4.1   Authority and duties

The contract contains specific provisions regarding the extent of the architect's authority and duties. These provisions are discussed in detail below. It would be quite wrong to think, however, that the provisions specifically set out in the contract are the end of the matter. The contract is very brief and, even in the case of a much more comprehensive contract form such as SBC, the architect has obligations which are sensible but not always immediately apparent.

Contractor's remedies

From the contractor's point of view, the architect's authority is indeed defined by the contract. Thus an architect who attempts to exceed this authority may, quite rightly, be ignored by the contractor. In fact, if the contractor carries out an instruction which the contract does not authorise the architect to issue, the employer is under no legal obligation to pay for the results. Indeed, the contractor would be in breach of contract. In that kind of case the contractor cannot take any legal action against the architect under the contract, to which the architect is not a party, though possibly the architect may find that the contractor has grounds for direct redress. Of course the contractor's chances of success will depend on all the circumstances. In view of the fact that the contractor would have no contractual remedy through the main contract, the courts might be prepared to consider an action in tort.

In order to prove negligence, three criteria must be satisfied:

1. There must be a legal duty of care; *and*
2. There must be a breach of that duty; *and*
3. The breach must have caused damage.

To take an example, if the architect instructed the contractor to carry out work on land owned by the employer but outside the site boundaries as shown on the contract drawings, the architect would be acting outside any express or implied authority of the contract. The contractor would be foolish to carry out such an instruction because it would leave itself open to a legal action by the employer

The JCT Minor Works Building Contracts 2016, Fifth Edition. David Chappell.
© 2018 John Wiley & Sons Ltd. Published 2018 by John Wiley & Sons Ltd.

for trespass at the very least. The architect would possibly be on the receiving end of an action from the employer under the architect's terms of engagement.

It is possible that the contractor could take direct legal action against the architect. The architect, in any event, will still be liable to third parties for negligent statements if it can be shown that the third party acted in reliance on the advice and the architect knew of that reliance and accepted it. The architect will have liability to the client on this basis which will run together with the liability under the conditions of engagement. Indeed, an architect's tortious liabilities may exceed those contained in the conditions of engagement. The architect may be liable to the contractor for duties other than statements. Actions by contractors against architects based on reliance on negligent drawings, specifications and possibly contract administration or even tender information, are a possibility.

Architect's appointment

The architect's powers and duties (see Tables 4.1 and 4.2) flow directly from the agreement with the employer. It is always prudent for them to enter into a formal written contract and the Codes of Conduct of both the RIBA and the ARB require the architect to do so. The RIBA Domestic Project Agreement 2010 (2012 Revision) or the RIBA Concise Agreement 2010 (2012 Revision) are suitable appointment documents to use with MW and MWD. The architect's appointment document will determine the precise extent of the architect's authority, but it is important to remember that, whatever the terms of the appointment may be, it has no effect on the building contract between contractor and employer. Therefore, provided that the exercise of the architect's authority is within the limits laid down in the building contract, the contractor may safely carry out the architect's instructions, take notice of certificates, etc. without worrying whether the architect has obtained the employer's consent.

For example, the Concise Agreement states that the architect must not make any material amendment to the approved design without the employer's consent. MW and MWD, however, empower the architect to issue instructions which may both alter the design and increase the total cost of the Works. The contractor must carry out such instructions and the employer is bound to pay. If the employer did not consent to the alteration in design or the increased expenditure, the employer's remedy lies against the architect.

Employer's responsibility

If the architect fails to carry out any duties under the contract, it may be held to be a default for which the employer will be liable to the contractor. But the employer cannot become liable until he or she is aware of the need to remind the architect of an obligation. The employer is not liable for what the architect does or fails to do in the capacity as certifier, because the parties have given the architect authority to form and express opinions. However, the employer is responsible for controlling the architect on becoming aware that the architect is not acting in accordance with the contract.

Table 4.1 Architect's powers under MW and MWD.

Clause	Power	Comment
2.1.2 (MWD only)	Direct the contractor in regard to the integration of the CDP.	The directions must not affect the design of the CDP without the contractor's consent under clause 3.4.2.
2.1.2 (2.2.1)	Be reasonably satisfied with the quality of materials or the standard of workmanship.	If they are stated to be a matter for the architect's approval.
2.10 (2.11 under MWD)	Instruct contractor not to make good defects.	An appropriate deduction may be made from the Contract Sum.
3.3.1	Consent in writing to sub-contracting.	Sub-contracting does not affect the contractor's duties.
3.4	Issue written instructions.	The contractor must comply forthwith.
3.5	Require the contractor to comply with an instruction by serving written notice.	The contractor has seven days from receipt of the written notice in which to comply. If it fails to do so the employer may employ and pay others to carry out the work and an appropriate deduction may be made from the Contract Sum.
3.6	Order an addition to or omission from or other change in the Works or the order or manner in which they are to be carried out or (in the case of MWD) a change in the Employer's Requirements.	The variation is to be valued by the architect on a fair and reasonable basis unless its value is agreed with the contractor before the instruction is carried out.
3.6	Agree the price of variations with the contractor prior to the contractor carrying out the instruction.	
3.8	Issue instructions requiring the exclusion from the Works of any person employed thereon.	The power must not be exercised unreasonably or vexatiously.
4.5.5	Issue a pay less notice on behalf of the employer	If the employer gives notice to the contractor that the architect is so authorised.
6.4.1	Give notice to the contractor specifying a default.	The contractor has seven days in which to end the default.

Table 4.2 Architect's duties under MW and MWD.

Clause	Duty	Comment
2.3 (2.4 under MWD)	Issue any further information necessary for the proper carrying out of the Works and issue all certificates.	See Table 4.3.
2.4 (2.5.1 under MWD)	Correct inconsistencies in or between the contract drawings, specification and schedules.	If the correction results in a change it is to be treated as a variation under clause 3.6.1.
2.5.2 (MWD only)	Approve the manner in which the contractor proposes to deal with an inconsistency.	If there is an inconsistency in or between documents prepared by the contractor for the CDP Works.
2.7 (2.8 under MWD)	Make in writing such extension of time as may be reasonable and notify the parties.	If it becomes apparent that the Works will not be completed by the stated completion date and the causes of delay are beyond the control of the contractor including compliance with an architect's instruction not due to the contractor's default and the contractor has so notified the architect.
2.9 (2.10 under MWD)	Certify the date when the Works have reached practical completion and the contractor has complied sufficiently with clause 3.9.	Clause 3.9 requires the contractor to comply with the CDM Regulations.
2.10 (2.11 under MWD)	Notify the contractor of defects in the Works within 14 days after the expiry of the rectification period.	The defects must be due to materials, goods or workmanship not in accordance with the contract.
2.11 (2.12 under MWD)	Certify the date when the contractor has discharged its obligations in respect of defects liability.	The contractor's obligation is limited to remedying defects, shrinkages or other faults appearing within the rectification period.
3.4	Confirm oral instructions in writing.	There is no time limit.
3.6.3	Value variation instructions	The value is to be on a fair and reasonable basis and where relevant the prices in the priced document must be used. Any direct loss and/or expense incurred by the contractor must be included if due to regular progress affected by compliance with instructions.

Table 4.2   (Continued)

Clause	Duty	Comment
3.7	Issue instructions as to expenditure of provisional sums.	If the price is not agreed, the instruction is to be valued as a variation under clause 3.6.3.
4.3	Issue interim certificates to the contractor.	Certificates are to be issued within five days after each due date and each due date is seven days after each interim valuation date.
4.8.2	Issue a final certificate	Not later than five days after the due date.
5.4A.2.3	Issue reinstatement work certificates.	To pay insurance money to the contractor if the contractor is responsible for effecting the Works insurance policy.
Schedule 3 Para 3.3	Issue instruction to confirm a change proposed by the contractor.	If the employer wishes to implement the change. The instruction must refer to the change, any adjustment to the Contract Sum, the contractor's share of financial benefit and any adjustment to the date for completion.

Reasonable skill and care

In performing professional services, the architect will be expected to act with the same degree of skill and care as the average competent architect. If the architect professes to have skills or experience which are greater than average, judgment may be stricter accordingly. Thus, if architects hold themselves out as experts in particular types of buildings, they will be expected to show a higher degree of skill in that particular area. If something goes wrong, to say that fellow architects without that specialisation would have taken the same action will be insufficient defence. It would have to be shown that a body of specialised opinion would have come to the same conclusion. The body need not be substantial in numbers provided that it exists and is responsible.

Knowledge of the law

This is not the place to discuss in detail architects' general duties to their clients and to third parties. One aspect of those duties, however, does affect administration of the contract. That is the duty to the client to be familiar with those aspects of the law which affect architectural work. An architect is not expected to have the detailed knowledge of a specialist building contracts lawyer, but must be capable of advising the client regarding the most suitable contract for a particular project. This implies that the architect must have, at least, a working knowledge of the main forms of contract. In addition, the architect must be able to give sensible advice on the choice of contract. Although problems may arise during the currency of the contract which clearly demand the attentions of a legal expert, the architect's client will be less than impressed and may well consider an architect to be incompetent who is unable to explain the basic provisions of the contract and deal with the day-to-day running of the job without legal assistance. If the client is put to unnecessary expense due to the fact that the architect had inadequate knowledge of the contractual provisions, the client may take legal action against the architect to recover any losses. The architect is also expected to be aware of decisions of the courts relevant to architecture and construction.

50

Agency

Up to the moment the employer and contractor enter into the contract, the architect has been acting for the employer as agent with limited authority. It is important that the architect understands what agency means. To say that someone is an agent means that person has the authority to do certain things on behalf of another person. That other person is termed the principal. Although I use the word 'person,' an agent or principal could be limited companies of course. The extent of an architect's authority to act on behalf of the employer or client will be set out in the architect's terms of engagement. It may be useful to highlight some general points which apply to all agents. An agency relationship may be formed in four ways:

- *Expressly*: This is when the client specifically appoints the architect either in writing or orally. A written appointment is not only the best method of appointment, it also is required by the Codes of Conduct of both the RIBA and ARB. It gives little scope for misunderstandings.
- *By implication*: This is when it is obvious to other people that the architect is acting as agent. The employer may behave as if the architect was acting as agent or the architect may simply be doing the kind of things that an agent would do.
- *By necessity*: The architect may possibly act for the employer in an emergency in a way that is outside the terms of engagement. This will be a rare occurrence. There would have to be a real emergency otherwise one can imagine a situation where the architect has acted and subsequently the employer does not agree. There could be considerable argument over whether the architect actually did act as agent or was just an interfering busybody.
- *By ratification*: This is when the architect does something on behalf of the employer which the architect has no authority to do but which the employer subsequently sanctions. There are two conditions: (i) the architect must be carrying out the action on behalf of the employer; and (ii) the employer must have been able to carry out the action when it was carried out by the architect.

An agent can have two kinds of authority:

- *Actual*: This is the kind of authority given by the terms of engagement which clearly set out what the architect may do.
- *Ostensible*: This is the authority the architect seems to have so far as people other than the employer are concerned. Often a contractor will simply assume that if the architect says something during the construction process, the employer has authorised it. Both architects and contractors must be careful about this. If a building contract is in use which sets out what the architect is authorised to do, the contractor will be entitled to treat such things as being actions for which the architect is authorised by the employer. The truth may be that the terms of engagement put particular limits on what the architect may do. The most common situation is when the terms of engagement say that the architect may not issue an instruction for a variation without permission from the employer. If the architect issues that kind of instruction without the employer's permission, the contractor will be entitled to carry it out and be paid for it by the employer because that is what the building contract says. The architect may have to pay the cost.

These are the duties of every agent:

- To act when action is required;
- To obey the principal's reasonable and lawful instructions;
- To exercise the reasonable skill and care required from that profession;
- Not to take a secret bribe or profit;
- To declare a conflict of interest;
- Not to delegate without authority;
- To keep proper accounts.

4.2 Duty to act fairly

During the contract the architect is expected to continue to act as the employer's agent, but also to administer the terms of the contract fairly between the parties. The employer may find the change difficult to understand and, to avoid problems, the wise architect will explain the situation, perhaps at the same time as sending the contract documents for signature (Box 4.1). Many years ago, it used to be thought that the architect was in a quasi-arbitral role (i.e. acting like an arbitrator) and, therefore, immune from any action for negligence by either party. That was never the case and certainly it is not the position today.

The architect may be in the position of deciding his or her own default, e.g. in considering an extension of time. It has been suggested that an architect owes a duty of care to a contractor when the contract calls on the architect to act fairly between the parties. Such things as extensions of time and certification of money would fall under this heading. It is an interesting idea but one which does not currently find much support.

> **Box 4.1   Letter from architect to employer explaining the duty to act impartially.**
>
> Dear Sir
>
> PROJECT TITLE
>
> [*Insert the main point of the letter then continue:*]
>
> This is probably the right moment to explain the additional responsibility which the Building Contract imposes upon me as the work progresses.
>
> Until the Building Contract is signed I am required, under the terms of my Engagement, to act solely as your agent within the limits set out in the Engagement. Thereafter, although I continue to act for you as before, I have the additional duty of administering the terms of the Building Contract fairly between the parties. In effect, this means that I am obliged to make any decisions (such as extensions of time, the amounts in interim certificates and the content and dates of other certificates) under the Building Contract strictly in accordance with its terms. If you require any further explanation of the position, I would be delighted to meet you.
>
> Yours faithfully

If the employer suffers loss or damage due to the fact that the architect has issued information late and the architect has granted an extension on that basis, the employer may seek to recover any loss from the architect. It might be thought that the architect's position is slightly less hazardous under this contract than it is under SBC or IC because of the absence of a full provision enabling the contractor to claim 'loss' and 'expense' under the contract as a result of architect's defaults, e.g. through late supply of information. However, if the contractor brought a successful action against the employer at common law for breach of contract, the employer might well be able to recoup the loss from the architect. Even if no such action were brought, the employer might sue the architect for loss of liquidated damages.

4.3   An architect in a local authority or similar

The extent of the authority and duties of an employee of the employer, for example in local or central government or in the construction department of a large organisation, is not always very clear. If the architect is in this position, the contractor may well consider that the architect at certain times is acting as agent for the employer and that any instruction issued which is not expressly empowered by the contract is, in effect, a direct instruction from the employer. The dangers of incurring unexpected costs are obvious and the architect should ensure that the extent of authority is made clear to the contractor at the beginning of the contract (Box 4.2).

> **Box 4.2   Letter from architect to contractor if architect is employee in local authority, etc.**
>
> Dear Sir
>
> PROJECT TITLE
>
> Possession of the site will be given to you by the Employer on the [*insert date*] in accordance with the Contract provisions.
>
> In order to avoid any misunderstandings, I should make clear, as far as you are concerned as the Contractor, that the extent of my authority is set out in the Contract. Although I am an employee of [*insert name of Employer*], I have no general power of agency to bind the Employer outside the express Contract terms. If I ever have cause to write to you on behalf of the Employer, I will clearly so state. If any matters are to be decided by me in accordance with the terms of the Contract, I will make such decisions impartially between the parties.
>
> Yours faithfully

As architect employee, the duty to act fairly between the parties remains, but it is admittedly difficult for the architect to convince the contractor that he or she is acting impartially. The administration of the contract under these circumstances calls for complete integrity, not only on the part of the architect but also on the part of the employer who has a duty to ensure that the architect carries out professional duties properly in accordance with the contract. Although it is usual for the actions of local authority employees to be governed by Standing Orders of the Council, they are of no concern to the contractor unless they have been specifically drawn to its attention at the time of tender. Thus an order that only chief officers may sign financial certificates would not avail the local authority as a defence if the architect was unable to issue a certificate because the chief officer was unavailable. In those circumstances the local authority could be liable for the architect's failure to issue a certificate. The contractor is entitled to rely on the architect's apparent authority. Orders that certificates may not be issued without authority from the audit department are likewise irrelevant so far as the contractor is concerned unless they are made part of the contract.

4.4   Express provisions of the contract

Article 3 of MW provides for the insertion of the name of the architect. The person so named will be the person referred to in the conditions whenever the word 'architect' appears. It is common practice to insert the name of the architectural

Box 4.3   Letter from architect to contractor naming authorised representatives.

Dear Sir

PROJECT TITLE

This is to inform you formally that the Architect's authorised representatives for all the purposes of the Contract are:

[*insert name*] - Director/Partner in charge of administering the Contract.

[*insert name*] – Project Architect.

Until further notice the above are the only people who are authorised to exercise the authority of the Architect in connection with the Contract.

Yours faithfully

For and on behalf of [*insert the name of the architect/practice in the contract*]

Copies: Employer

Quantity Surveyor

Consultants

Clerk of works

firm (or the name of the chief architect in a local authority) because of the difficulties which could arise in having to renominate every time the project architect left the firm, died or retired. It is seldom the cause of any dispute, but it is prudent to notify all interested parties of the name of the authorised representative, i.e. the project architect, who will administer the contract on a day-to-day basis (Box 4.3). Changes in the identity of the project architect should be notified in the same way. The employer is under a duty to appoint another architect if, for example, the named architect resigns, retires or dies, and Article 3 makes this duty explicit.

55

Signing

It is not strictly necessary for the architect named in Article 3 to sign all letters, notices, certificates and instructions personally, but they must be signed by a person properly authorised. It is not wise for the authorised representative to sign his or her name only, even if using headed stationery, because the letter, certificate, etc. must be signed by or in the name of the architect named in the

56   contract. Otherwise the letter may be taken to be written merely on the authority of the person who signs, with serious financial consequences. The architect should not use a rubber stamp. It is not good practice although it is probably valid. Signing someone else's name is wrong although common. To sign in someone's name is best accomplished by procuration ('per pro' or 'pp'). For example: pp John Smith (then underneath the signature of the person actually signing the letter).

Access to the works and premises

There is no express provision in MW or MWD giving the architect and representatives access to the Works at all reasonable times. Such a right of access to the site is implied under the general law. However, there is a gap, because the architect may need access to the contractor's workshops etc., where items are being prepared for the contract. If the architect does need to visit the contractor's premises to keep an eye on things, there must be an appropriate clause in the contract documents, following the wording of SBC clause 3.1, because it is probable that the architect does not have that right of access to either the contractor's or any sub-contractor's workshops under the general law.

Architect ceasing to act

If the named architect dies or ceases to act for some other reason, the contract states that it is the employer's duty to nominate a successor within 14 days of *Article 3*   the death or ceasing to act of the named architect. Unlike the position under SBC, there is no provision for the contractor to object to the successor, but there is nothing to prevent the contractor from referring the matter to immediate adjudication or arbitration if it strongly objects about the replacement. However, unless the identity of the replacement prevents it, the contractor must continue with the Works while awaiting the outcome of the dispute resolution process. In the majority of contracts carried out under this form, it is possible that the Works will be complete by the time a decision is reached. That is probably the reason for the exclusion of the right of objection in the first place. Clearly, the employer would do well to listen carefully if the contractor makes any representations, in order to avoid such difficulties. It is regrettable that an employer will occasionally try to take over the former architect's role personally. An employer is not entitled to do that, but must 57   appoint another architect to the job.

There is an important proviso which states that 'no replacement appointee as Architect…shall be entitled to disregard or overrule any certificate, opinion, decision, approval or instruction given by' the former architect unless the for- *Article 3*   mer architect would have had the power to do so. This is a sensible provision to safeguard the contractor's interests under circumstances which are in the sole control of the employer. A successor architect who disagrees with some of the predecessor's decisions should notify the employer immediately (Box 4.4), to safeguard the position. The proviso clearly cannot mean that certificates or instructions given by the former architect must stand for all time, even if they

---

Box 4.4   Letter from architect to employer if disagreeing with former architect's decisions, etc.

Dear Sir

PROJECT TITLE

I have now had the opportunity to examine all the drawings, files and other papers relating to this Contract and I have visited site and spoken to the Contractor.

Article 3 of the Contract prevents me from disregarding or overruling any certificate or instruction given by the previous architect. This does not prevent me from issuing further certificates and, particularly, further instructions which may vary instructions given by the previous architect. Of course in the latter instance there will be a cost implication.

I list below the matters on which I find myself unable to agree entirely with the decisions of the previous architect:

[list all matters subject to disagreement]

Of course, other areas of disagreement may come to light as the Contract progresses. In the meantime, I suggest that you consider these matters and then we should meet to discuss the best ways of dealing with them.

Yours faithfully

58

are wrong. The implication of the proviso is that if it is necessary for the successor architect to make changes, they may, if appropriate, be treated as variations and the contractor will be entitled to payment accordingly. For example, if the former architect had given instructions for the construction of a detail which, in the opinion of the new architect, would lead to trouble, the new architect could issue further instructions correcting the matter and the contractor would be paid for correcting the work. Unlike instructions, certificates are expressions of the architect's opinion in tangible form for the purposes specified in the contract. Certificates of practical completion and making good are not, in any event, susceptible to change. On the other hand, an architect is normally entitled to issue a financial certificate which is for a lower gross sum than previously certified and, being cumulative, it is difficult to see how a new architect could forfeit this right. Indeed, in a case where the former architect had wrongly certified a sum due to the contractor, the replacement architect would have a duty when issuing the next certificate to correct the error. Table 4.3 shows all the certificates to be issued by the architect.

Table 4.3   Certificates to be issued by the architect under MW and MWD.

Clause	Certificate
2.3 (2.4 under MWD)	General duty to issue certificates
2.9 (2.10 under MWD)	Practical completion
2.11 (2.12 under MWD)	Making good
4.3	Interim certificates
4.8.2	Final certificate
5.6.5.1	Reinstatement works certificates

Further information

The amount, if any, of further information necessary will vary greatly depending on the size and complexity of the work. In the case of small simple projects, the contract drawings may need very little elaboration. Whether the information is necessary should be a matter of fact, not opinion. The contractor is expected to use its own practical experience in carrying out the work. Note that there is no obligation on the contractor to apply for further information either in this clause or anywhere else in the contract. The prudent contractor will do so, but it is the architect's duty to supply correct information to enable the contractor to carry out the Works properly in accordance with the contract. The date for completion is part of the provisions of the contract, therefore the architect's duty is to supply necessary information at the correct times to enable the contractor to proceed with the Works and to complete them by the contract date for completion. If the architect fails to do so, it will be a breach for which the contractor can claim extension of time and possibly damages at common law. Of course, a contractor who merely stands by while the architect fails to issue information at the right time is unlikely to attract much sympathy from an adjudicator or an arbitrator. It is always wise for a contractor to ask for information it requires, allowing a reasonable period for the architect to respond.

The issue of certificates refers not only to interim payments but also to such things as practical completion and making good after the rectification period. Although there is no particular way in which a certificate should be set out, it is a formal document. It may be in the form of a letter, but to avoid any doubt it is always good practice to head the letter 'Certificate of…' and begin 'I certify…'. If a number of certificates are to be issued in sequence it is normal to number them in order. Delay in the issue of a certificate is a serious matter and may give rise to financial claims by the contractor (Table 15.1).

Confirming instructions will be dealt with in Chapter 16.

The remaining express duties of the architect are covered in the appropriate chapters of this book. Because this contract is so brief, the duties are also briefly stated. This can be misleading and the architect should always bear in mind that the courts will imply terms to cover the way in which the architect must administer the contract. Generally, the architect is expected to act promptly and efficiently. An architect who does so will avoid claims and contribute towards the smooth running of the contract.

4.5   Common problems

Correspondence between architect and contractor is usually by email nowadays. To what extent is such correspondence binding?

1.6.1
61

the contract makes clear that all notices, instructions or other communications which are referred to in the contract, must be in writing. Communications sent by fax or email are 'in writing', because they can easily be printed out. If the contract does not specifically say how the communication is to be sent (for example, termination correspondence can only be sent in one of the specified ways), it can be sent 'by any effective means'. Both email and fax are effective means of communication. It is sufficient to conclude important negotiations by email and online signature even though the signature was simply 'Guy'. The courts recognise that nowadays business is transacted by email some of which is very informal in nature. So far as the signature is concerned, it can be an initial, first name or even a nickname (presumably if everyone knew who that was) any of which would suffice as the signature. E-mail is often used to form contracts and contracts formed in this way will be enforceable.

62
63

It is common for an architect to put two rectification periods in the contract: one for building work and the other, usually longer, for mechanical and electrical. Is this allowed?

4.8.1

The straight answer is 'no'. The reason for the different periods is that it is said that heating and the like need to be tested over a while year while building work generally can be tested for defects over three months (the default period in these contracts) or six months. Unfortunately for that theory, the contract only provides for one period. There is only one certificate of making good and the issue of the final certificate is linked with it. The solution of course is to make the rectification period the longer of the two being considered.

5 Contractor's powers and duties

5.1 Contractor's obligations: express and implied

The contractor's powers and duties are set out in Tables 5.1 and 5.2 respectively.

5.2 Basic principles

It is all too commonly assumed that the whole of the contractor's contractual obligations are set out in the printed contract form. This is not so and this particular contract form is silent on many important matters. These gaps are filled in by terms which the law will take as being part of the contract and which are referred to as 'implied terms' (see Chapter 3).

Some implied terms

If there was nothing at all in the contract about the contractor's obligations, the general law would require the contractor to do three things:

- To carry out its work in a good and workmanlike manner exercising reasonable care and skill. This means that the contractor must show the same degree of competence as the average contractor experienced in carrying out that type of work.
- To use materials of good quality which are reasonably fit for their purpose.
- To ensure that the completed building or structure is reasonably fit for its intended purpose provided that purpose is known to the contractor when it signs the contract. This obligation is modified where the employer engages an architect to design the Works, because the architect is then responsible for the design.

A term would be implied that the contractor would comply with the building regulations and with other statutory requirements – a matter which is specifically referred to in MW and MWD.

2.1

Table 5.1 Contractor's powers under MW and MWD.

Clause	Power	Comment
2.2 (2.3 under MWD)	Commence the Works on the specified date.	
3.1	Assign the contract.	If the employer gives written consent
3.3.1	Sublet the works or part thereof.	If the architect consents in writing.
3.4.2 (MWD only)	Consent to an instruction affecting the design of the CDP Works.	
4.4.1	Make application to the architect.	Stating the sum the contractor considers due at the relevant due date and the basis of calculation.
4.4.2.2	Give a payment notice to the architect.	If an interim certificate is not issued at the right time. The notice must state the sum the contractor considers due at the relevant due date and the basis of calculation.
4.7.1	Suspend performance of obligations under the contract	If employer fails to pay, and has not issued effective pay less notice, by final date for payment. Contractor must first give seven days' notice and must resume work after payment in full
4.7.2	Recover a reasonable amount in costs and expenses.	Incurred by the contractor as a result of its suspending performance.
5.5	Reasonably require evidence from the employer that the insurance referred to in clauses 5.4B or 5.4C has been taken out and is in force.	
5.7	Terminate the contractor's employment by notice if just and equitable.	Given within 28 days of the occurrence of loss or damage.
6.8.1	Give notice to the employer specifying the default.	The notice must be served on the employer by special or recorded signed-for delivery or delivery by hand Notice can be served: ■ If the employer fails to pay by the final date for payment the amount due plus any VAT *or* ■ If the employer interferes with or obstructs the issue of any certificate *or* ■ If the employer fails to comply with the requirements of the CDM Regulations.

Table 5.1   (Continued)

Clause	Power	Comment
6.8.2	Give notice to the employer.	The notice must be served on the employer by special or recorded signed-for delivery or delivery by hand Notice may be served if the whole or substantially the whole of the Works is suspended for one month due to: ■ architect's instructions; or ■ impediment, prevention or default by the employer, architect, etc.
6.8.3	Terminate the contractor's employment.	If default or suspension not ended seven days after notice Termination takes effect on receipt of this notice The notice must not be given unreasonably or vexatiously.
6.9	Terminate the contractor's employment.	By notice if the employer is insolvent as defined in clause 6.1 The notice does not require a prior default notice and it takes effect on receipt by the employer.
6.10	Give notice to the employer that unless the suspension ends within seven days, the contractor's employment may be terminated.	The notice must be served on the employer by special or recorded signed-for delivery or delivery by hand Notice may be served if the whole or substantially the whole of the Works is suspended for one month due to: ■ *force majeure; or* ■ architect's instructions as a result of negligence or default of statutory undertaker; *or* ■ loss by Works insurance or excepted risks; *or* ■ civil commotion or terrorism; *or* ■ exercise of UK or local government power.
	Terminate the contractor's employment under the contract.	By notice served on the employer by special, recorded signed-for delivery or actual delivery if the suspension does not end.

(Continued)

Table 5.1   (Continued)

Clause	Power	Comment
7.1	Agree to resolve the dispute by mediation.	
Schedule 1 para 1	By written notice jointly with the employer to the arbitrator stating that they wish the arbitration to be conducted in accordance with any amendments to the JCT 2016 CIMAR.	
Schedule 1 para 2.1	Serve on the employer a written notice	If the contractor wishes a dispute to be resolved by arbitration.
Schedule 1 para 2.3	Give a further arbitration notice to the employer referring to any other dispute.	After the arbitrator has been appointed Rule 3.3 applies.
Schedule 3 para 3.1	Propose changes to the designs and specifications and programme.	If it benefits the employer.
Schedule 3 para 4.1	Suggest economically viable amendments.	If they may result in an improvement in environmental performance.

These implied terms can be modified by the express terms of the contract itself and are, in fact, modified by these contracts, with the result that the contractor is under a lesser duty than would otherwise be the case.

Some statutory terms

Statute also imposes implied obligations on the contractor and in particular the Supply of Goods and Services Act 1982 implies terms as to the quality of goods supplied by the contractor under the contract. The construction of dwellings (both houses and flats) is governed by the Defective Premises Act 1972 (in Northern Ireland, the Defective Premises (Northern Ireland) Order 1975) which provides:

'Any person taking on work for or in connection with the provision of a dwelling...owes a duty to see that the work that he takes on is done in a workmanlike manner, with proper materials...and so as to be fit for the purpose required...'.

The Act envisages a two-stage test. Therefore, if the dwelling is fit for habitation despite the fact that some work was not done in a workmanlike manner nor with proper materials, it seems that it complies with the Act.

64

Effect of sub-letting

The express provisions of MW and MWD must be read against this background. There is a further important point. Even if the contractor sublets the Works or any part of them with the architect's written consent, sub-contracting

Table 5.2 Contractor's duties under MW and MWD.

Clause	Duty	Comment
2.1	Carry out and complete the Works in accordance with the contract documents in a good and workmanlike manner and in accordance with statutory requirements and give all notices.	In accordance with the construction phase plan.
2.1.1 (MWD only)	Complete the design for the CDP using reasonable skill, care and diligence.	The contractor is not responsible for the Employer's Requirements or the adequacy of any design therein.
2.1.2 (MWD only)	Comply with CDM Regulations 8 to 10 and with the architect's directions for integration of the design of the CDP with the Works as a whole	
2.1.3 (2.2.2 under MWD)	Take all reasonable steps to encourage employees and agents, including sub-contractors, to be registered cardholders under the Construction Skills Certification Scheme.	
2.1.3 (MWD only)	Provide the architect with two copies of drawings, etc. to explain the CDP.	As and when necessary.
2.2 (2.3 under MWD)	Complete the Works by the specified date.	The architect must insert the date in the contract particulars.
2.5.1 (2.6.1 under MWD)	Immediately give to the architect a written notice specifying the divergence.	If the contractor finds a divergence between statutory requirements and the contract documents or architect's instructions.
2.5.2 (MWD only)	Correct any inconsistency in or between the CDP documents.	At the contractor's own expense.
2.6 (2.7 under MWD)	Pay all fees and charges in respect of the Works.	Provided they are legally recoverable.
2.7 (2.8 under MWD)	Notify the architect of delay.	If: ■ It becomes apparent that the Works will not be completed by the specified date *and* ■ This is because of reasons beyond the control of the contractor including compliance with any instruction of the architect whose issue is not due to a default of the contractor.

(Continued)

Table 5.2   (Continued)

Clause	Duty	Comment
2.8.1 (2.9.1 under MWD)	Pay liquidated damages to the employer.	If the Works are not completed by the specified date or by any later date fixed under clause 2.7 (2.8 under MWD).
2.10 (2.11 under MWD)	Make good at own cost any defects, shrinkages or other faults.	The defects, shrinkages or other faults must have appeared within the rectification period *and* must be due to materials or workmanship not in accordance with the contract *and* unless the architect has instructed otherwise.
3.2	Keep a competent person in charge on the site at all reasonable times.	Under clause 3.8 the architect may instruct this person's exclusion.
3.4	Carry out all architect's instructions forthwith	The instructions must be in writing.
3.9	Comply with the CDM Regulations.	
4.8.1	Supply the architect with all documentation reasonably required to enable the final sum to be computed.	■ Within the specified period of the date of practical completion ■ It is not conditional upon a request from the architect.
4.6.1	Pay simple interest to the employer on amount not properly paid in the final certificate, until paid.	If the contractor fails to pay any amount due to be paid by the final date for payment.
5.1	Indemnify the employer against any expense, liability, loss, claim or proceedings in respect of personal injury or death.	If the expense, etc. arises out of or in the course of or is caused by reason of the carrying out of the Works except to the extent that it is due to any act or neglect of the employer, employer's persons or any statutory undertaker.
5.2	Indemnify the employer against and insure and cause its sub-contractors to insure against any expense, liability, loss, claim or proceedings for damage to property other than the Works or site materials.	This is subject to clauses 5.2.1 to 5.2.3. The indemnity operates if the expense etc. ■ Arises out of or in the course of or is caused by reason of the carrying out of the Works *and* ■ to the extent that it is due to any negligence, omission or default of the contractor or any contractor's person.

Table 5.2 (Continued)

Clause	Duty	Comment
5.3	Take out and maintain and cause its sub-contractors to maintain the insurances necessary to meet its liability under clauses 5.1, 5.2, 5.3.1 and 5.3.2 including compliance with all relevant legislation.	
5.4A	Insure against all risks.	New Works only.
5.4C	Insure as stated in the contract.	If the contract states that clause 5.4C applies.
5.5	Produce evidence of insurances.	If the employer reasonably so requires.
5.6.1	Forthwith notify the architect and the employer.	If there is loss or damage to the Works or due to an excepted risk or to existing structure or contents.
5.6.3	Authorise payment of insurance money to employer.	Employer may retain amount for professional fees.
5.6.4	Restore damages work, replace materials, remove and dispose of debris and proceed to carry out and complete the Works.	After any inspection required by the insurers.
5.7	Terminate the contractor's employment.	Within 28 days of loss or damage if just and equitable.
6.7.1	Give up possession of the site.	Where the employer terminates the contractor's employment
6.11.2	Prepare an account setting out: ■ Total value of work properly executed and materials properly brought on site and other amounts due and ■ Cost to the contractor of moving from site *and* ■ Direct loss and/or damage	Upon termination of the employment of the contractor under clauses 6.8 to 6.10. The employer must pay the full amount properly due within 28 days of submission of the account
Schedule 3 para 1	Work with the employer in a co-operative and collaborative manner, in good faith and a spirit of trust and respect. Support collaborative behaviour and address behaviour not collaborative	
Schedule 1 para 2.1	Endeavour to establish environment where health and safety is paramount.	

(Continued)

Table 5.2   (Continued)

Clause	Duty	Comment
Schedule 3 para 2.2	Comply with approved codes of practice from the Health and Safety Executive. Ensure personnel have proper health and safety training. Ensure personnel have access to health and safety advice. Ensure full and proper consultation with personnel.	In addition to contractual health and safety requirements. In accordance with Regulation 7 of the Health and Safety (Consultation with Employees) Regulations 1996. In accordance with the Health and Safety (Consultation with Employees) Regulations 1996.
Schedule 3 para 3.2	Provide details of proposed changes with assessment of financial benefit to employer and a quotation.	
Schedule 3 para 3.3	Negotiate with employer to agree value of change, financial benefit and new date for completion.	If the employer wishes to implement the contractor's proposed change.
Schedule 3 para 4.2	Provide all information reasonably requested by the employer.	Regarding environmental impact of materials and goods.
Schedule 3 para 5.2	Provide all information reasonably requested by the employer.	For monitoring contractor's performance.
Schedule 3 para 5.3	Submit proposals for improving performance.	If employer thinks targets may not be met.
Schedule 3 para 6	Promptly notify employer of any matter likely to cause a dispute and senior executives must meet to try to resolve the matter.	
Schedule 3 para 8.1	Include in any sub-contract the requirements of Regulation 113(2)(c)(i) and (ii)	Where the employer is a local or public authority and Regulation 113 of the Public Contracts Regulations 2015 applies.
Schedule 3 para 8.2	Include in any sub-contract certain specified provisions.	Where the employer is a local or public authority and the Public Contracts Regulations 2015 apply.
Schedule 3 para 8.3.1	Include in any sub-contract provisions enabling the contractor to terminate the sub-contractor's employment under Regulation 57.	Where the employer is a local or public authority and the Public Contracts Regulations 2015 apply.
Schedule 3 para 8.3.2	Take appropriate steps to terminate a sub-contractor's employment and, if required by the employer, appoint a replacement sub-contractor.	Where the employer is a local or public authority and the Public Contracts Regulations 2015 apply and so require.

3.3.1

does not free the contractor from responsibility for the sub-contracted work. The contractor remains responsible to the employer for all defaults of the sub-contractor as regards workmanship, materials or otherwise.

Summary of powers and duties

Tables 5.1 and 5.2 summarise the contractor's powers and duties respectively under the express provisions of MW and MWD.

5.3 Carrying out the Works

2.1

The contract says that the contractor must carry out and complete the Works in accordance with a number of criteria:

- In a proper and workmanlike manner, and
- In compliance with the contract documents, and
- In compliance with the construction phase plan, and
- In accordance with statutory requirements.

In MWD there are additional duties related to the CDP Works which are considered in Chapter 22. There is no express requirement that the contractor must proceed regularly and diligently. That point is discussed below.

This is probably the most important part of the contract. It sets out the contractor's basic obligation. The contractor is bound to complete the Works by the date for completion set out in the contract particulars. The Works must be brought to a state where they are practically completed so that the architect can

2.9 (2.10)   issue the certificate of practical completion.

Qualifications to the obligations

The contractor's basic obligation is subject to two qualifications:

6.8

- The contract provisions which entitle the contractor to terminate its employment under the contract, e.g. if the employer is in default as stated in the contract or becomes insolvent.
- If the employer – or the architect – prevents the contractor from completing its work by the completion date unless, of course, a proper extension of time has been validly made.

Contract documents

Second recital (third recital)

The contractor must carry out its work in compliance with the contract documents. The architect must take care to ensure that the description of the work is clear and comprehensive and this means using precise language. All too often important parts of the contract documents are written in slovenly English. It is impossible to enforce generalisations. The contract documents must contain all the requirements which the employer wishes to impose, and the use of phrases such as 'quality' or 'of durable standard' should be avoided. It is regrettable that there is increasing use of phrases like 'quality materials' which are quite meaningless unless the word 'quality' is properly qualified. Moreover, the contract makes clear that the content of other contract documents cannot override or modify the printed conditions; the effectiveness of this provision has been upheld by the courts on many occasions.

The contractor must complete *all* the work shown in, described by or referred to in the contract documents. The contractor's obligation only comes to an end when the architect has issued the practical completion certificate. Thereafter, the contractor must fulfil its obligations in respect of any defects arising during the rectification period (see Chapter 20).

There is no further amplification of the contractor's basic obligation although the contractor is, of course, bound to complete the Works *by* the date stated in the contract particulars and if it does not do so, subject to the operation of the provisions for extension of time, it must pay liquidated damages to the employer. The contractor is entitled to complete early, but the architect is not obliged to provide information at times to suit the contractor's proposed early completion.

Proceeding regularly and diligently

There is no express requirement for the contractor to proceed regularly and diligently. The only references to diligence is in MWD referring to the contractor's obligation to complete the design of the CDP Works and in the contractor's obligation to restore work with due diligence after loss or damage. This is surprising, particularly in view of the termination provisions, which permit the employer to terminate the contractor's employment if it fails to proceed regularly and diligently. Whether a term that the contractor must proceed regularly and diligently would be implied in this instance is open to question. However, the employer's ability to terminate on this ground emphasises the importance of understanding its meaning. The phrase 'regularly and diligently' has been defined by the Court of Appeal as follows:

Although the contractor must proceed both regularly and diligently with the Works, and although each word imports into that obligation certain discreet concepts which would not otherwise inform it, there is a measure of overlap between them and it is thus unhelpful to seek to define two quite separate and distinct obligations.

What particularly is supplied by the word 'regularly' is not least a requirement to attend for work on a regular daily basis with sufficient in the way of men, materials and plant to have the physical capacity to progress the work substantially in accordance with the contractual obligations.

What in particular the word 'diligently' contributes to the concept is the need to apply that physical capacity industriously and efficiently towards the same end.

Taken together the obligation upon the contractor is essentially to proceed continuously, industriously and efficiently with appropriate physical resources so as to progress the works steadily towards completion substantially in accordance with the contractual requirements as to time, sequence and quality of work.

Beyond that I think it impossible to give useful guidance. These are after all plain English words and in reality the failure of which [the clause] speaks is, like the elephant, far easier to recognise than to describe'.

68

Whether a contractor is or is not proceeding regularly and diligently clearly depends on all the circumstances, but it is something which most architects instinctively recognise.

Programme

The sensible architect will require a contractor to provide a programme, preferably showing logic links, but mere failure to comply with a programme is not usually sufficient alone to show failure to proceed regularly and diligently. The architect will not approve the contractor's programme, because it is essentially a setting out of the contractor's intentions. In any event, it is not thought that approval by the architect has any great significance in this context. It certainly does not make the architect responsible for the correctness of the programme.

69

5.4   Workmanship and materials

The contractor must use materials and workmanship to the quality specified in the contract documents or, if so stated, in accordance with the architect's approval. The contract states that if approval of workmanship or materials is stated to be a matter for the architect's opinion, then quality and standards must be to the architect's reasonable satisfaction. They must be satisfactory to the architect on the basis of an objective standard informed by the concepts of honesty, good faith and genuineness and the absence of arbitrariness, capriciousness, perversity and irrationality. MWD expressly deals with the situation where materials and workmanship are not specified nor made subject to the architect's approval or satisfaction In such instances, the standard of materials and workmanship are to be appropriate to the Works. If the CDP is concerned, the standard is to be appropriate to the CDP or to the Works. It is not clear why MW omits this 'appropriate' provision in regard to the Works. Nevertheless, it is omitted from MW and, therefore, in the absence of clear description in the contract documents and if the quality is not stated to be to the architect's approval the contractor is probably entitled to provide materials of a reasonable or adequate standard.

This provision in fact imposes an obligation on the architect to define the quality and standards of workmanship and materials very carefully indeed. The employer normally places reliance on the architect, not the contractor, to specify correctly and the contractor is unlikely to be liable if it supplies the wrong material merely because the architect has wrongly indicated that any one of a range of materials may be used. Where it is impossible adequately to specify in

2.1

2.1.2 (2.2.1)

70

(2.2.1)

71

sufficient detail, the architect could do worse than fall back on the phraseology of the common law:

- All work shall be carried out in a good and workmanlike manner and with proper care and skill.
- All goods and materials shall be of good quality and reasonably fit for their intended purpose.

Quite clearly, the contractor is expected to show a reasonable degree of competence and to employ skilled tradesmen. Although the architect has no power to direct the contractor as to how it should carry out its work, save to the extent that the architect can by a variation instruction change the manner in which the Works are to be carried out, the architect does have power to instruct that any unsatisfactory employee or anyone else employed on the Works should be excluded.

3.6.1

3.8

The contractor *must* provide workmanship, materials and goods of the standards and quality specified. This is a matter which is at the contractor's risk and failure to supply is a breach of contract. Having said that, there is an argument that, if a specified item becomes unavailable, the contract might be frustrated. Being unavailable does not mean being more difficult or more expensive to obtain.

Certification

The architect is *bound* to include in progress certificates the value of any materials and goods which have been properly brought to the site and reasonably for the purposes of carrying out the Works, provided that they are adequately protected from the weather or other causes of damage. Use of the words 'properly' and 'reasonably' is presumably to prevent the contractor bringing materials onto site far in advance of the date they are required simply to secure valuation and payment. The architect would be entitled to ignore prematurely delivered materials in any certificate. Indeed, the architect would have no power under the contract to certify the value of such materials.

4.3.2

Who owns the materials?

This is a very dangerous provision from the employer's point of view since the unfixed materials and goods will not necessarily become the property of the employer, even though the employer has paid for them, if, in fact, they are not the employer's property in law, as will often be the case. Builders' merchants frequently include a 'retention of title' clause in their contracts of sale (see Chapter 18). Ideally, this provision should be amended so as to ensure that the inclusion of the value of unfixed materials is a matter for the architect's discretion. Less satisfactorily, it may include an appropriate provision requiring the contractor's application for progress payments to be accompanied by documentary proof of ownership.

It is probable that the architect is safe from an action for negligence for operating the contract as it stands because the employer has signed MW or MWD which says that the architect *shall* include on-site unfixed materials. However, architects should be alert to the dangers, and the Joint Contracts Tribunal should amend this provision. This plea was first made in the first edition of this book, in 1990 – clearly to no avail.

5.5   Statutory obligations

2.1.1 (2.1)

2.6 (2.7)

The contract imposes on the contractor a duty to comply with all statutory obligations, e.g. those imposed by the Building Regulations and the CDM Regulations and subsequent amendments and give all notices they require. The contractor is deemed to have allowed for the payment of all fees and charges which are legally 'demandable' under any statutory requirement. The contractor must pay these fees even if it has omitted to include for them in its tender price.

2.5.1 (2.6.1)

The contractor has an obligation to give immediate written notice to the architect if the contractor discovers any divergence between the statutory requirements and the contract documents or one of the architect's instructions. There then follows a provision which expressly excludes the contractor from any liability under the contract if the Works do not comply with statutory requirements. There is an important proviso that the contractor must not have been in breach of its obligations to immediately notify the architect if any divergence was found and that the non-compliance must have resulted from carrying out the Works in accordance with the contract documents or in accordance with an architect's instruction. This exclusion does not apply to CDP Works.

72

The effect of this provision is contractually to exempt the contractor from liability to the employer if the Works do not comply with statutory requirements provided the contractor has carried out the work in accordance with the contract documents or any architect's instruction if, for example, the contractor does not spot the divergence. The contractor's obligation to notify the architect of a divergence only arises *if* the contractor spots it, and unless the contractor does so, it is under no obligation to notify the architect. Plainly, it cannot exonerate the contractor from the duty to comply with statutory requirements and if, for example, the contractor carried out work in breach of the Building Regulations, it would be criminally liable and open to prosecution, because the primary liability to comply with them rests on the contractor. The wording is probably enough to protect the contractor from any action by the employer. But it does not protect the architect of course, and if the fault is the architect's the employer will be able to recover any losses from the negligent architect.

5.6   Contractor's representative

3.2

The contractor must keep a 'competent person in charge' on the Works at all reasonable times, i.e. during normal working hours. This person is intended to be the contractor's full-time representative on site, but the appointment and replacement of the person in charge are not subject to the architect's approval although in an appropriate case the architect could exercise powers to require the exclusion of the person in charge from the site and thus force the contractor's hand.

Competent means that the contractor's representative must on a fair assessment of the task and all the factors involved be regarded by the contractor as competent to perform. It is essential that the architect knows this person's identity from the outset, because he or she is the contractor's agent for the purpose of accepting instructions. Instructions given to the person in charge are *deemed* to be given to the contractor. From the contractor's point of view, such a person must be completely reliable.

Many architects include in the tender documents a requirement that the contractor give adequate notice of the replacement of the person in charge, and this is good contract practice.

5.7   Compliance with architect's instructions

The contractor must carry out the architect's instructions forthwith and this obligation is not conditioned in any way. Except under MWD, the contractor is given no right to object to architect's instructions, even those involving a variation. Under MWD, the contractor's consent is required to any instruction affecting the design of CDP work.

If a contractor fails to comply with an architect's instruction, the architect may serve the contractor with a written notice requiring compliance. If the contractor has not complied within seven days from receipt of that notice, the employer may engage others to carry out the work and an appropriate deduction is to be made from the Contract Sum. Box 16.4 is a suggested letter from the architect, and Box 16.5 is a possible reply.

To avoid the possibility of an argument that the employer has waived its rights, it is essential that the architect ensures that the contractual provisions are put into operation if the contractor fails to comply with instructions.

MW and MWD contain no express provision for the architect to issue instructions for the removal of work not in accordance with the contract, but the broad wording of the contract must extend to cover that situation.

Other rights and obligations

Tables 5.1 and 5.2 summarise the contractor's rights and duties generally. Other matters referred to in those tables are dealt with in the appropriate chapters.

5.8   Suspension of obligations

The contractor has the right to suspend performance of its obligations under the contract under certain circumstances. In effect, this means that the contractor may cease work on site and elsewhere. In order to be able to do so, certain criteria must be satisfied. First, the employer must have failed to pay money due by the final date for payment. Second, the contractor must have given seven days' notice in writing of its intention. The contractor must resume work as soon as it is paid in full. There is no ordinary common law right to suspend work for late payment although, in practice, it is often done. A contractor who suspends work legitimately will be entitled to a reasonable amount in respect of costs and expenses incurred as a result of exercising the right to suspend. In order to get the costs and expenses, the contractor must apply to the architect with whatever information the architect needs to ascertain the amount to be paid. To 'ascertain' means to 'find out for certain' or to 'get to know'. Therefore, the architect cannot simply make a rough estimate. Although the contract says nothing, there is no doubt that the contractor is also entitled to have the period of suspension disregarded in computing the contractual time for carrying out the Works. The date for completion must be adjusted. It is arguable that the extension of time provisions in the contract do not give the architect power to extend time in these circumstances because it may be difficult to argue that the suspension was beyond the contractor's control. Nevertheless, statute is clear that the date must be adjusted, possibly by the employer.

5.9 Common problems

The contractor has left site before practical completion and there is no sign that it will return so the architect issues an instruction requiring the contractor to return to site.

This appears to be an exercise in futility. The contract requires the contractor to carry out and complete the Works. If it is not on site, the contractor cannot do that. In order to complete the Works, the contractor must be on site and working. Suspension of the Works and failure to work regularly and diligently are grounds for the employer to terminate the contractor's employment. Therefore, the architect's instruction is an instruction to the contractor to do something which it should be doing anyway. There may occasionally be a time for that sort of instruction, but it will be very rare. The architect should, with the agreement of the employer, issue a default notice to warn the contractor that if it does not return to site within seven days, the employer may terminate the contractor's employment (see Chapter 21). That tells the contractor that the employer will not stand for any kind of hit and miss attendance and if the contractor persists, it opens the door to allow the employer to get another contractor to finish the Works and claim any extra costs from the original contractor.

The adjudication provisions and the contractor's right to suspend have been deleted from the contract. The employer has not paid the last interim certificate. What can the contractor do?

If the Works are not to the employer's own house, statute says that the Scheme will apply. That will effectively reinstate the deleted provisions. If it is

6.8.1.1

the employer's own house, all is not lost because the contractor can always serve a default notice under the termination provisions. If the employer still refuses to pay, the contractor may terminate its employment and the employer will then be liable, not only for the unpaid interim certificate amount, but also for any damages suffered by the contractor as a result of the termination. The main damages are likely to be loss of the profit which the contractor would have made by completing the project. That could be a considerable amount. There is interest on top, of course. The contractor should beware of simply leaving site if not paid, because that would be a breach of contract.

6   Employer's powers and duties

6.1 Powers and duties: in the contract and elsewhere

Some of the employer's powers and duties are clear from the MW and MWD contracts, but some arise under the general law.

Tables 6.1 and 6.2 show those powers and duties of the employer which are set out in the contract.

A few words about obligations under the general law (implied terms)

These are provisions which the law writes into a contract in order to make it commercially effective (see also Chapter 3). Terms will be implied to the extent that they are not inconsistent with the express terms which may exclude or modify the implied terms.

It is an implied term of MW and MWD that the employer will do all that is reasonably necessary to bring about completion of the contract. Conversely, it is implied that the employer will not so act as to prevent the contractor from completing in the time and in the manner envisaged by the agreement. If the employer is in breach of either of these implied terms and the contractor suffers a loss, the contractor can recover the loss against the employer through one of the dispute resolution procedures.

If the employer – either personally or through the agency of the architect or that of anyone else for whom the employer is responsible – hinders or prevents the contractor from completing the Works by the date for completion, not only is the employer in breach of contract, but he or she will be disentitled from enforcing the liquidated damages clause.

Briefly

The duty has been put in different ways, but in essence the position can be summarised as follows:

- The employer and the employer's agents must do all things necessary to enable the contractor to carry out and complete the Works expeditiously and in accordance with the contract.

The JCT Minor Works Building Contracts 2016, Fifth Edition. David Chappell.
© 2018 John Wiley & Sons Ltd. Published 2018 by John Wiley & Sons Ltd.

Table 6.1 Employer's powers under MW and MWD.

Clause	Power	Comment
2.8.2 (2.9.2 under MWD)	Deduct liquidated damages from any monies due to the contractor under this contract or recover them as a debt.	If the Works are not completed by the completion date or any extended date and provided that a prior notice and pay less notice has been given.
3.1	Assign the contract.	If the contractor consents.
3.5	Employ and pay others to carry out the work.	If the contractor fails to comply with a written notice from the architect requiring compliance with an instruction within seven days from its receipt. An appropriate deduction is to be made from the Contract Sum.
4.5.4	Give written pay less notice to the contractor.	The notice must state the sum considered to be due, the basis of calculation and must be given not less than five days before the final date for payment of an interim or final certificate or payment notice.
4.5.5	Give written notice to the contractor.	If the employer wishes to authorise someone other than the architect to give a pay less notice.
5.5	Reasonably require the contractor to produce evidence of insurance.	If the contractor is required under the contract to effect and maintain insurance under clauses 5.3, 5.4A, 5.4B or 5.4C or if the contractor is responsible for ensuring the policy is effected and maintained.
5.6.5.2	Retain amounts properly incurred and notified to insurers in respect of professional fees.	Where the contractor is responsible for effecting Works insurance policy and employer must pay insurance monies to contractor.
5.7	Terminate the contractor's employment.	Within 28 days of any loss or damage if it is just and equitable to do so.
6.3.2	Reinstate the contractor's employment.	On terms agreed by the parties.
6.4.2	By written notice terminate the contractor's employment under the contract.	Within 10 days if the contractor has not ended the specified default within seven days of receiving a default notice.
6.5.1	By written notice terminate the contractor's employment under the contract.	At any time if the contractor is insolvent.
6.5.2.3	Take reasonable measures to ensure that the site, the Works and materials on site are protected and not removed.	The contractor must allow and not hinder the measures.
6.6	By written notice terminate the contractor's employment under this or any other contract.	If the contractor is guilty of corruption or the circumstances in Regulation 73(1)(b) of the Public Contracts Regulations 2015.

Table 6.1   (Continued)

Clause	Power	Comment
6.7.1	Engage others to carry out and complete the Works. With others, take possession of the site. Use all temporary buildings, etc.	After the employer has terminated under clauses 6.4 to 6.6.
6.10.1	Give written notice warning that if suspension is not ended within seven days, termination may follow.	The notice must be served on the contractor by special, recorded signed-for delivery or by hand. Notice may be served if the whole or substantially the whole of the Works is suspended for one month due to: ■ *force majeure; or* ■ architect's instructions as a result of negligence or default of statutory undertaker; *or* ■ loss by risk covered by Works insurance or by excepted risk; *or* ■ civil commotion or terrorism; *or* ■ exercise of UK or local government power.
	Terminate the contractor's employment under the contract.	By notice served on the contractor by special, recorded signed-for or actual delivery if the suspension does not end
Schedule 1 para 1	By written notice jointly with the contractor to the arbitrator stating that they wish the arbitration to be conducted in accordance with any amendments to the JCT 2016 CIMAR.	
Schedule 1 para 2.1	Serve on the contractor a written notice.	If the employer wishes a dispute to be resolved by arbitration.
Schedule 1 para 2.3	Give a further arbitration notice to the contractor referring to any other dispute.	After the arbitrator has been appointed Rule 3.3 applies.
Schedule 3 Para 5.3	Inform the contractor that a target for a performance indicator may not be met.	If the employer considers that to be the case.
Schedule 3 Para 8.3.2	Require the contractor to terminate the sub-contractor's employment in accordance with Regulation 71(9). Require the contractor to appoint a replacement sub-contractor.	The contractor must take appropriate steps.

Table 6.2 Employer's duties under MW and MWD.

Clause	Duty	Comment
2.8.3 (2.9.3 under MWD)	Inform the contractor in writing if there is an intention to deduct liquidated damages from the final certificate sum.	Not later than the issue of the final certificate.
3.5	Employ and pay others to carry out work necessary to comply with the architect's instruction.	If the contractor does not comply with an instruction within seven days after receiving a compliance notice from the architect.
3.9	Comply with the CDM Regulations.	
3.9.1	Ensure that the principal designer carries out his or her duties. Ensure that the principal contractor carries outs its duties.	If the contractor is not the principal contractor.
3.9.4	Immediately notify the contractor with details.	If the employer appoints a replacement for the principal designer or principal contractor.
4.1	Pay to the contractor any VAT properly chargeable.	
4.5.1	Pay to the contractor amounts certified by the architect	Payment must be made by the final date for payment subject to any pay less notice.
4.5.2	Pay to the contractor amounts stated in the contractor's payment notice.	If the certificate is not properly issued and the contractor's payment notice has been given. Payment must be made by the final date for payment subject to any pay less notice.
4.5.4	Give the contractor a pay less notice.	If the employer intends to pay less than the sum stated as due in a certificate of payment notice. The pay less notice must be given not later than five days before the final date for payment.
4.6	Pay simple interest at 5% above Bank of England official rate to contractor on amount not properly paid, until paid in full.	If employer fails to pay any amount due by final date for payment.
5.4B.1	Maintain joint names policy against the specified perils.	For existing structures and contents until the date of issue of the certificate of practical completion or termination if earlier.
5.4B.2	Maintain joint names policy against all risks.	For the Works until the date of issue of the certificate of practical completion or termination if earlier.

Table 6.2 (Continued)

Clause	Duty	Comment
5.4C	Insure as stated in the contract.	If the contract states that clause 5.4C applies.
5.5	Produce evidence of insurances.	If the contractor reasonably so requires.
5.6.5.1	Pay all monies from insurance which the contractor has taken out to the contractor in instalments	If the contractor is responsible for Works insurance policy.
6.7.3	Set out an account in a statement.	Within a reasonable time after completion of the Works and making good of defects if the employer has terminated and the architect has not set out an account in a certificate.
6.11.4	Pay to the contractor the full amount properly due in respect of the contractor's account.	If the contractor's employment is terminated under clauses 6.8 to 6.10. Payment must be made within 28 days of submission of the account by the contractor.
Schedule 3 para 1	Work with the contractor in a co-operative and collaborative manner, in good faith and a spirit of trust and respect. Support collaborative behaviour and address behaviour not collaborative.	
Schedule 3 para 2.1	Endeavour to establish environment where health and safety is paramount.	
Schedule 3 Para 3.3	Negotiate with contractor to agree value of change, financial benefit and new date for completion.	If the employer wishes to implement the contractor's proposed change.
Schedule 3 para 5.1	Monitor and assess the contractor's performance.	By reference to any performance indicators in the contract documents.

- Neither the employer nor the employer's agents will in any way hinder or prevent the contractor from carrying out and completing the Works expeditiously and in accordance with the contract.

More detail

The scope of these implied terms is very broad and more and more claims for breach of them have come before adjudicators and the courts. The employer must not, for example, attempt to give direct orders to the contractor, and must see that the site is available to the contractor on the date specified in the con-
2.2 (2.3)   tract and that access to it is unimpeded by the employer's persons. For example, in one instance, squatters had occupied part of the site and as a result the

80   employers were unable to give possession on the due date. This was a breach of contract which caused appreciable delay to the contractors who were entitled to damages. Availability is especially important in the case of work to existing structures or occupied buildings, and these are matters which should be discussed between the employer and the professional advisers before the contract is let (see Chapter 11 for an explanation of 'Employer's Persons' and 'Contractor's Persons').

6.2   Rights under MW and MWD

Who are the parties?

The architect is sometimes referred to as a party to the contract and some architects even express concern that the contractor may seek adjudication against them. Of course the architect is not a party to the contract. The confusion sometimes arises because the architect is named in the contract, but that does not make the architect a party. Therefore, the contractor cannot seek adjudication against the architect because adjudication is only available to parties to the contract. The contract is between the employer and the contractor, who are the only parties to it. Although a party, the employer has few express rights of any substance.

The employer's major and most important right is to have the Works properly completed in accordance with the contract documents by the agreed completion date. The employer also has the right to assign the benefits of the

3.1   contract to a third party, provided the contractor consents in writing.

Damages for non-completion

If the contractor does not complete the Works by the agreed completion date, the employer is entitled to recover liquidated damages at the rate specified in the contract particulars for each complete week, day or other specified period during which the Works remain uncompleted after the original or extended completion date.

There is no requirement that the architect should issue a certificate of non-completion, although it is good practice for the architect to notify the employer in writing when the completion date has passed, a copy can be sent to the contractor (a standard form is available for this purpose). This reminds both parties that liquidated damages are now payable. The mere fact of late completion

2.8 (2.9)   is sufficient to bring the liquidated damages provision into operation. The employer is given the right to deduct liquidated damages from monies due to the contractor, but if (unusually) no sums are due or are to become due to the contractor, the employer must take action to recover the damages as a debt.

For some reason, best known to the JCT, the contract is not crystal clear about what the employer needs to do in order to recover liquidated damages. It is important, because if the employer fails to issue some notice which some adjudicator, arbitrator or court subsequently decides is essential, the employer may be deprived of liquidated damages.

In order to be sure of recovery, it is suggested that the employer does the following although it is recognised that some of what follows may be unnecessary. The employer should:

1. Check with the architect that the date for completion has passed and that no extension of time (or further extension of time) is due to be issued; and

2.8 (2.9)   2. Write to the contractor referring to the relevant clause and state that the contractor is required to pay or allow liquidated damages at the rate stated in the contract particulars between the date for completion (as extended if appropriate) and the date of practical completion (it is wise to issue this letter (notice) as early as possible but it must be issued at latest before the
2.8.3 (2.9.3)   date of issue of the final certificate); and

3. No later than five days before the final date for payment (of the certificate or payment notice from which the employer intends to deduct the liquidated damages) give the contractor a pay less notice stating the sum the employer considers to be due on the date of the notice and the way in which the sum has been calculated. This notice must include details of how the liquidated damages have been calculated. For example, it should be sufficient if the employer refers to the calculation of the sum in the architect's certificate or in the contractor's payment notice, shows how the liquidated damages have been calculated (e.g. number of weeks x rate per week) and then shows the deduction of the damages from the calculation to produce the sum considered to be due in the pay less notice.

In practice, the architect should, of course, advise the employer of all rights before the contract overruns. Waiting until the final certificate to deduct liquidated damages may not be a good idea.

6.3   Other rights

These are summarised in Table 6.1 and are there described as 'powers'. They are discussed in the appropriate chapters.

6.4   Duties under MW and MWD

Failure to carry out a duty

The essence of a duty is that it is something that must be done. It is not something that the employer can chose whether or not to do; it is mandatory. Breach of a duty imposed by the contract will make the employer liable to an action for damages by the contractor for any loss the contractor can prove.

Some breaches of contract may, in fact, entitle the contractor to treat the contract as at an end. They are called 'repudiatory breaches' which means that they go to the basis of the contract (see Chapter 21). For example, physically expelling the contractor and its operatives from site would be a repudiatory breach because the employer is effectively demonstrating a wish no longer to be bound by the contract. Alternatively, for the contractor to walk off site permanently before the Works are complete would be a repudiatory breach on its part.

Breach of any contractual duty on the part of the employer will always, in theory, entitle the contractor to at least nominal damages, therefore, the employer must be careful. In some cases of course, any loss will be difficult if not impossible to quantify.

Payment

From the contractor's point of view, the most fundamental duty of the employer is to make payment in accordance with the terms of the contract. However, while steady payment against certificates is essential from the contractor's viewpoint, the general law does not usually regard non-payment or late payment as a major breach of contract. Certainly non-payment of one certified sum to the contractor is not generally regarded as repudiation. Chapter 18 deals in detail with payment.

81

Time for payment

MW and MWD are quite specific in their terms as to payment (see Chapter 18). It is essential that the employer is made aware of the need to pay promptly in accordance with the contract terms because nothing sours good working relationships more than late payment, and most contractors do have a cash-flow problem. In these circumstances the architect ought to write to the employer pointing out the contractual position (Box 6.1).

It cannot be stressed too much that payment does not depend on when the architect decides to issue a certificate. The architect must issue a certificate within five days of the due date. In these contracts the due date is seven days after the interim valuation date referred to in the contract particulars. Payment must be made within 14 days of the due date unless the contractor is entitled to issue a payment notice when it will be extended (see Chapter 18, section 18.4), which means payment before the expiry of that period. This is a very tight timetable. Even if the architect issues the certificate immediately after the due date, the employer only has 13 days in which the receive it and pay. If the architect does not issue the certificate until the fifth day after the due date, the employer will have barely ten days to pay. Delivery of the certificate by hand to the employer or, if that is not practicable, by 'next day' special delivery is the next best thing. These days, there is no reason why the certificate cannot be e-mailed to the employer immediately it is issued and a copy put in the post.

Box 6.1   Letter from architect to employer if employer is slow to honour certificates.

Dear Sir

PROJECT TITLE

The contractor has complained to me that payments are not being made until after the final date for payment.

Under the terms of the contract you have 14 days from the due date within which to make payment. These key dates are on each interim certificate in the top right hand corner. In addition, I always notify you of the last date for payment when I issue a certificate.

Failure to pay by the final date for payment entitles the contractor to suspend performance of all its obligations under the contract or to terminate its employment under the contract. If the contractor did either of these things the consequences could be very expensive. In addition, sums outstanding entitle the contractor to interest at 5% above Bank of England official bank rate.

Quite apart from the strict legal requirements, it is good practice to pay promptly, because the contractor is always in the position of having paid out substantial sums well before payment is due. Prompt payment is essential to cash flow and to good working relations.

If you would bear this in mind, there should be benefits on all sides.

Yours faithfully

Payment notices

82

The payment position has been enormously complicated by the introduction of payment notices. If the architect does not issue an interim certificate by the end of the 5th day after the due date and the contractor has already issued an application for payment, that application becomes a payment notice, or, if the contractor has not issued an application, it may issue a payment notice. In the absence of an interim certificate, the employer must pay the amount on the payment notice. Needless to say, it is part of the architect's duty to advise

83

the employer about all these complex payment matters.

Paying less

If the employer wishes to pay less than the amount on the certificate or payment notice a pay less notice must be issued no later than five days before the final date for payment. If no pay less notice is issued, the employer must pay the amount in the interim certificate or, if no interim certificate, in the payment notice. Even if it seems obvious that work has not been carried out properly and the project is littered with serious defects, payment must be made. The simple reason is because the payment provisions are only to decide who holds the money until everything is worked out. They are not final. The advantage lies with the employer, who has the power to issue a pay less notice, to decide how much to pay. But if the employer does not use that advantage, the money may be in the contractor's hands even if it amounts to more than the work that has been done at that point. More details about pay less notice are in Chapter 18.

How to pay

Payment by cheque is probably good payment, but it is no excuse for the employer to say, for example, that the computer arrangements do not fit in with the scheme of certificates, which is an increasingly common excuse for late payment. If this is indeed the case, the payment period, or better still the computer arrangements, should have been amended before the contract was let. In the absence of a pay less notice, there is no excuse for failure to pay.

Some employers believe that their obligation to pay does not start until they receive an invoice from the contractor. On the other hand, some contractors believe that they can require the employer to pay by submitting an invoice even where there is no interim certificate. Both beliefs are wrong.

6.5   Retention

The employer has certain rights in the retention percentage, which is commonly 5%. The percentage of work to be valued and certified up to practical completion is to be inserted in the contract particulars. The default entries are 95% and $97^1/_2$% respectively. It is not stated that the retention is trust money, which would have to be kept in a separate account, and it is likely, therefore, that the contractor is at risk if the employer becomes insolvent.

The employer's rights in the retention are to have it as a fund from which defects may be remedied and other bona fide claims settled. But morally the retention is the contractor's money and once again it is suggested that it would be an improvement if the JCT would amend the contract to provide that the employer's interest in the retention was 'fiduciary as trustee for the contractor'.

6.6   Other duties

These are summarised in Table 6.2 and are commented on as appropriate in other chapters.

6.7   Common problems

In order to keep everything 'friendly' an employer may tell the contractor that liquidated damages will not be deducted.

To be blunt, this kind of statement is sometimes nothing but a badly considered lie. It is surprising how many employers make this statement but later decide to take the damages anyway because they are larger than they anticipated or because, when making the statement, the employer did not realise that the delay was going to be so long. Although liquidated damages are intended simply to compensate the employer for any delay due to the fault of the contractor and not a punishment for late completion, it cannot be denied that if a contractor is told that no liquidated damages will be deducted, it removes one of the pressures. An employer who promises not to deduct liquidated damages may be unable to change his or her mind later. That will be the case if, relying on the promise, the contractor acts in a certain way which will cost it money if the employer has a change of mind. The most common example is if the contractor, relying on the promise, does not deduct damages from the sub-contractor which actually caused the delay. In that case, the employer may be stuck with the promise. In legal jargon one would say that the employer was estopped.

85

The employer wants to dispense with the architect and act as contract administrator.

Unfortunately, this situation is relatively common; certainly where the MW is concerned. An employer will see the position entirely in terms of economics: why pay an architect's fees when the contractor is doing all the work. The employer overlooks the fact that a certificate from an architect stands as an independent certificate and carries a lot of weight in any legal proceedings, whereas if the employer issues a certificate it carries no more weight than the contractor's opinion. The architect's duty under the contract is to act fairly between the parties when certifying or making a judgment, for example, about the standard of workmanship. If the employer really wants to act as contract administrator, there appears to be nothing to stop that happening provided that the contractor know about it before submitting a tender. Once the contractor's tender has been accepted, the employer cannot unilaterally decide to be the contract administrator unless the contractor agrees.

86

87

7 Quantity surveyor

7.1 Appointment

MW and MWD make no provision for the appointment of a quantity surveyor. Nevertheless, the architect may wish to have the assistance of a quantity surveyor in valuing the Works.

Whether to appoint

The decision whether or not to appoint a quantity surveyor will be taken by the employer with the architect's advice. The advice will naturally take into account the amount of work which the quantity surveyor could be asked to carry out. In the case of small works, the use of a quantity surveyor tends to be the exception. The architect should be competent to deal with interim payments, the computation of the final sum and valuation of variations if the work is of a simple and straightforward nature, particularly if a clause is inserted to provide for stage payments. If, however, the architect has any doubts about competence in this field or the work is complex, the use of a quantity surveyor is indicated. It is important to remember that, if architects hold themselves out to their clients as capable of carrying out quantity surveying functions, that is the standard which will be expected of them. The architect should not put forward the architect's name as quantity surveyor unless the position has been checked with the professional indemnity insurers. Since there is no mention of the quantity surveyor in the contract, there appears to be no valid reason why the architect's name should ever be put forward. There is no benefit in so doing, and there may be insurance problems.

Early appointment

If the architect decides that a quantity surveyor is required for the Works, the employer should be informed at the earliest possible opportunity, no later than Stage 1 – Preparation and Brief, so that the quantity surveyor can assist throughout the project. The architect should take care to explain to the employer why a quantity surveyor is required (Box 7.1). The situation will be simplified if the employer has been given a copy of the RIBA Domestic Project Agreement 2010

The JCT Minor Works Building Contracts 2016, Fifth Edition. David Chappell.
© 2018 John Wiley & Sons Ltd. Published 2018 by John Wiley & Sons Ltd.

> **Box 7.1   Letter from architect to employer if the services of a quantity surveyor are required.**
>
> Dear Sir
>
> PROJECT TITLE
>
> I strongly advise you that the services of a quantity surveyor will be required for this project. A quantity surveyor is the specialist in building economics and will deal with the preparation of valuations for variations and interim certificates and provide overall cost advice. It is important that a quantity surveyor should be appointed at this stage so that you can derive the maximum benefit from the advice and the quantity surveyor can become fully involved with the project.
>
> Although I cannot recommend any particular quantity surveyor, you may wish to consider the use of [*insert name*] of [*insert address*], with whom I have worked many times in the past. If you will let me have your agreement, I will carry out some preliminary negotiations and advise you regarding the letter of appointment.
>
> If you require further advice about this matter it may be best for us to meet.
>
> Yours faithfully

(2012 Revision) or the RIBA Concise Agreement 2010 (2012 Revision) on one of which, presumably, the architect's engagement is based. Where the client is a consumer, a common situation where the main contract is MW or MWD, care must be taken to explain, individually negotiate and record each term whichever standard conditions of engagement are used. Otherwise, the terms may be invalid under the Unfair Contract Terms in Consumer Contracts Regulations 1999.

7.2 Duties

Because the contract does not provide for the appointment of a quantity surveyor, it is no surprise that there are no duties allocated to the quantity surveyor. If it is decided to make an appointment, it is sensible to include a clause in the contract to that effect. Although terms may be implied to cover the associated duties, it is preferable to deal with powers and duties of the quantity surveyor by means of a specially worded and inserted provision. There are two possible ways of dealing with this:

1. The insertion of a brief clause in the contract to cover the activities the architect intends the quantity surveyor to perform; or
2. A letter to the contractor informing him that the quantity surveyor is to be the architect's authorised representative in respect of specified activities.

A clause in the contract

The introduction of a clause in the contract is probably the more satisfactory way of accomplishing the objective. The contractor then knows, at tender stage, just what is intended. If the architect handles the matter by means of a letter at the beginning of the contract, the contractor might object violently to it and relations will have been soured at the start. Another danger is that the contractor may not appreciate the limits of the quantity surveyor's duties and may carry out as instructions what are simply the quantity surveyor's comments on some aspect of the work. Although the contractor would be wrong to do so, it is little consolation to the employer if disruption and delay results. If it is felt that the quantity surveyor's duties must be dealt with in this way, great care must be taken with the letter (Box 7.2).

If a clause is included, proper legal assistance must be obtained for its drafting. Among the duties that the architect wishes the quantity surveyor to carry out may be the following:

2.4 (2.5) ■ Valuation of the variation in the unlikely event that an inconsistency results in a variation.

2.10 (2.11) ■ Valuation of the amount of the employer's contribution under clause, if the architect instructs that the contractor should be paid for making good some defects.

2.10 (2.11) ■ Valuation of the appropriate deduction from the Contract Sum if the employer engages others to make good some or all of the defects.

■ Valuation of the appropriate deduction if the employer has to employ others to carry out work contained in architect's instructions.

3.5

3.6.3 ■ Valuation of variations.

3.7 ■ Valuation of instructions in connection with provisional sums.

4.3 ■ Measurement and valuations for the purpose of interim certificates.

■ Measurement and valuations for the purposes of the employer's pay less notice.

4.5.4

■ Calculation of a reasonable amount in respect of costs and expenses after the contractor's proper suspension.

4.7.2

4.8.1 ■ Computation of the final payment.

4.9 ■ Calculation of fluctuations.

5.6.4.1 ■ Calculation of instalments releasing insurance money.

6.7.3 ■ Valuations after termination by the employer.

6.11 ■ Checking the contractor's account prepared after termination.

Provision should be included for the quantity surveyor to require the contractor to supply any necessary documents for the purpose of carrying out the valuations, calculations or measurements. It is also prudent to stress that the quantity surveyor's duties cover quantification only with no power to agree or decide liability for payment on or to issue instructions, certification or to make any awards, and this would be the situation under the general law.

89

Box 7.2   Letter from architect to contractor regarding the duties of the quantity surveyor.

Dear Sir

PROJECT TITLE

This is to inform you that [*insert name*] of [*insert address*], who is a quantity surveyor, is appointed as my authorised representative for the purpose of carrying out the duties of valuation, calculation, checking, computation and measurement in respect of the following clauses of the contract:

2.4 [(2.5) *when using MWD*]
2.10 [(2.11) *when using MWD*]
3.5
3.6.3
3.7
4.3
4.5.4
4.7.2
4.8.1
4.9
5.6.4.1
6.7.3
6.11

You must supply all necessary documents, vouchers etc. to enable the quantity surveyor to carry out this work. Your attention is drawn to the fact that the quantity surveyor's duties are limited to quantification. The quantity surveyor is not empowered to issue instructions, certificates, decide extensions of time or decide liability for payment.

Yours faithfully

Copies: Employer
          Quantity Surveyor

7.3   Responsibilities

Assuming that the employer appoints a quantity surveyor, the quantity surveyor's responsibility is to the employer under the appointment. Even if a list of duties is included in a special clause inserted in MW and MWD, the quantity surveyor will not be liable to the contractor under that contract because, like the architect, the quantity surveyor is not a party to the contract.

Valuation

90

If the quantity surveyor makes a mistake in valuation, the architect must correct it. The onus on the architect to be satisfied about all the quantity surveyor's calculations so far as it is reasonable to do so cannot be emphasised too much. That does not mean that the architect must do a separate valuation; that would be to make the quantity surveyor redundant. However, the quantity surveyor must provide the architect with a breakdown of each valuation so that the architect is able to check that it generally conforms to the state of the work as the architect knows from inspections on site. It is of course essential that the architect must notify the quantity surveyor of any defective work or materials

91

so that they are excluded from any valuation. The quantity surveyor will be liable to the employer for any advice given to the employer or to the architect if the advice is negligently given. However, if the quantity surveyor makes a mistake, the architect will always be the first target for the client's displeasure.

Architect should not appoint

Architects should resist, as far as possible, any efforts on the part of the employer to get them to appoint quantity surveyors themselves. From the employer's point of view, it is understandable that it is preferable to deal with everything through one person, but the arrangement leaves the architect very exposed. The modern practice in negligence cases seems to be to sue everyone in sight and some commentators argue that it makes no real difference whether the employer appoints all consultants directly or simply appoints the architect and leaves the architect to appoint any other consultants, with the employer's permission, who may be necessary. That is thought to be a wrong view. Figure 7.1 indicates the various relationships which are possible. If the employer suggests that the architect appoint the quantity surveyor, it is suggested that the position should be placed on record (Box 7.3).

It should be remembered that the architect has no contractual relationship with the quantity surveyor (if the employer has appointed one); if there is any liability between architect and quantity surveyor it will be in tort.

7.4   Common problems

The quantity surveyor is engaged to carry out valuations and insists on sending them to the contractor at the same time as they are sent to the architect.

The contract does not provide for a quantity surveyor and, therefore, if a quantity surveyor is appointed, he should not be communicating with the contractor at all except perhaps to receive substantiation for variations and the like. It is quite common for a quantity surveyor under all the forms of contract to send valuations directly to the contractor as well as to the architect.

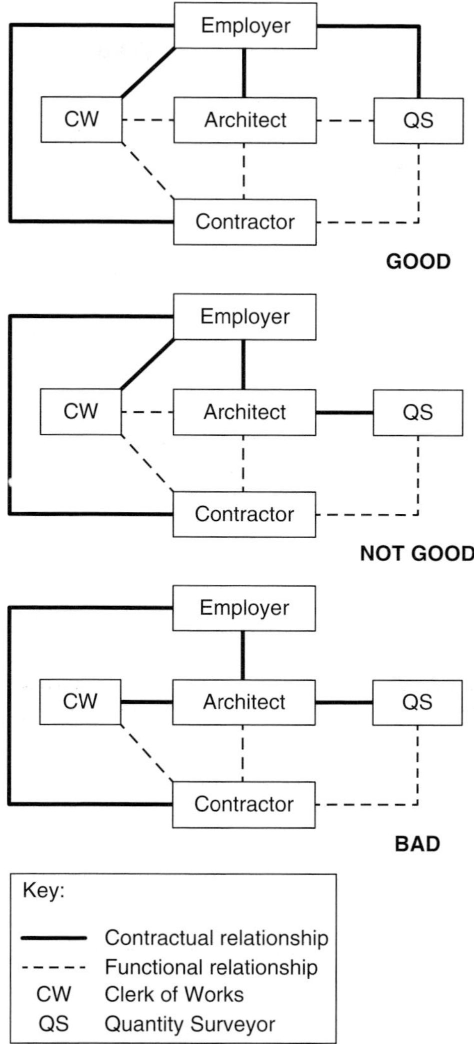

Figure 7.1   Some possible contractual relationships.

Not only is it quite wrong to do so, it has huge potential to start disputes. That is because the architect as certifier may adjust the quantity surveyor's valuation before certifying. For example, the architect should carry out an inspection just before certifying and may find defective work which must not be certified. Of course there may be other reasons why the architect certifies a figure which is different to the quantity surveyor's valuation. In these circumstances, a contractor armed with a valuation contrary to the certificate may try to argue that the architect has wrongly certified. Although the contract is clear about the architect's powers to certify, a deal of misunderstanding and bad feeling can be generated.

Box 7.3   Letter from architect to employer if he or she requires the architect to appoint the quantity surveyor.

Dear Sir

PROJECT TITLE

Thank you for your letter and I am pleased that you have agreed to the appointment of [*insert* name] as quantity surveyor for this project.

It is not appropriate for me to appoint the quantity surveyor and my professional indemnity insurance does not cover for such an appointment. In any event, the quantity surveyor is your consultant and provides the kind of cost information which is not available from an architect. Normal practice is for the employer to appoint all consultants.

I will, in any case, co-ordinate all professional work. If you wish me to do so, I can draft an appropriate letter of appointment for you to send.

Yours faithfully

Contractors often submit daywork sheets to support a claim for extra payment when carrying out variations. How should thy be taken into account by the architect or the quantity surveyor (if appointed)?

3.6.2

If the architect instructs a variation, the contract says that it must be valued by agreement with the contract or by the architect on a fair and reasonable basis using any prices in the specification or other document which the contractor has priced and which is one of the contract documents. If there are no relevant prices in the priced document, the architect must work out a price. Obviously, the architect can ask the contractor for substantiation and can ask for daywork sheets. However, if the architect decides to value the work using daywork sheets, it is essential to tell the contractor before the work is carried out so that the sheets can be filled in as the work is done. A problem with daywork sheets is that if the architect decides that is how the work is to be valued, the numbers of men, hours worked and materials used as recorded on the sheets cannot be challenged on the basis that the contractor could have done the work quicker. The arithmetic can be checked and whether the equipment and material were actually used, but that is all. A judge remarked that it is well known that when work is carried out on a daywork basis, it takes longer than would otherwise be the case!

92

If the contractor submits daywork sheets without being instructed to do so, the architect is entitled to ignore them.

8 Clerk of works

8.1 Appointment

MW and MWD make no provision for a clerk of works. This is because, quite clearly, on most Works for which MW and MWD are suitable, the employment of a clerk of works in a full or part-time capacity is hardly justified. Every project has its own difficulties, however, and if it is considered that a clerk of works is required, the architect must advise the employer accordingly.

The Domestic Project Agreement 2010 (2012 revision) and the Concise Agreement 2010 (2012 revision) both expressly state in the Small Project Services Schedule that the architect will '[V]isit the site to see that the work is proceeding generally in accordance with the contract'. If it seems that more frequent or even constant inspections are required, the architect should recommend the appointment of a clerk of works. This is something to consider before tenders are invited so that the presence of a clerk of works can be notified to the tenderers.

A clause in the contract

If a clerk of works is to be engaged, it is essential to include a clause to that effect in the contract. It is suggested that the clause follows the lines of clause 3.3 of the Intermediate Building Contact (IC). That provides that the employer is entitled to appoint a clerk of works, but that the duty of the clerk of works is simply to act as an inspector on the employer's behalf although under the architect's direction.

This states all that is necessary and removes any uncertainty in the contractor's mind regarding the directions of the clerk of works – none are empowered. If the clerk of works attempts to issue instructions, the contractor should write to the architect accordingly (Box 8.1).

If the employer is one of those organisations which employ clerks of works on their permanent staff, that is an excellent arrangement. To avoid misunderstandings, the position should be put on record (Box 8.2). In the case of most employers, however, it will be a one-off arrangement, the clerk of works being engaged by the employer on a full-time or part-time basis for the duration of

The JCT Minor Works Building Contracts 2016, Fifth Edition. David Chappell.
© 2018 John Wiley & Sons Ltd. Published 2018 by John Wiley & Sons Ltd.

Box 8.1   Letter from contractor to architect regarding directions from the clerk of works.

Dear Sir

PROJECT TITLE

The clerk of works has issued direction number *[insert number]* dated *[insert date]* on site, a copy of which is enclosed.

Such directions have, of course, no contractual effect. Clearly, the directions of the clerk of works issued in relation to the correction of defective work can be very helpful. We are anxious to avoid misunderstandings on site and in this spirit we suggest that the clerk of works should issue no further directions, other than those relating to defective work. There is no reason why all other matters cannot be referred directly to you by telephone and, if appropriate, you can issue an architect's instruction as empowered by the contract.

In our view, the above system would remove a good deal of the uncertainty which must result from the present state of affairs. We look forward to hearing your comments.

Yours faithfully

Box 8.2   Letter from architect to employer regarding the appointment of a clerk of works.

Dear Sir

PROJECT TITLE

In view of the nature/size/value *[delete as appropriate]* of this project, I advise the appointment of a clerk of works on a part-time basis, say *[insert number]* hours per week.

Where a clerk of works is appointed, it is for the Employer to appoint and pay that person under a separate agreement. A clerk of works will be able to make inspections of such frequency as should ensure the proper carrying out of the work. Such an appointment is likely to repay its cost several times over in savings on lost time and money as the contract progresses.

Although, naturally, I will carry out my own duties with reasonable skill and care, on a contract of this type where my duty is to inspect the work to see that it is proceeding generally in accordance with the contract, I cannot accept responsibility for such defects as would be discovered by the employment and consequently more detailed inspection of a clerk of works.

Yours faithfully

the contract. Some firms of architects employ their own clerks of works on a permanent basis, but that is not a good idea even though the employer is prepared to pay an additional sum to cover the clerk of works' salary (see below).

In the interests of the project, the architect should advise the employer to appoint the clerk of works as soon as the contractor's tender has been accepted. This gives the clerk of works time to become thoroughly familiar with the drawings, specification and schedules. The architect will hold a meeting with the clerk of works to brief the latter about the work to be done, but the architect would be wise to confirm the main points to the clerk of works by letter (Box 8.3), which should outline the aspects of the job considered to be important and act as a reminder of the extent and limitations of the clerk of works' duties.

8.2 Duties

If a clause is incorporated based upon the clause in IC, the clerk of works' sole duty is to inspect the Works. That clause makes clear that this function is carried out on behalf of the employer; the clerk of works is not the architect's inspector. This is vitally important as will be apparent shortly. The employer, however, has no power to direct the clerk of works. That is the architect's prerogative alone.

The architect's directions to the clerk of works will presumably embrace how, where and at what intervals inspections should be carried out and to what particular attention should be paid. If the architect issues any directions which are other than purely routine in nature, they should always be confirmed in writing to protect the architect's position. In practice, the clerk of works will often do more than simply inspect. The architect may seek the clerk of works' advice, as a person of some practical experience, from time to time. The contractor often asks the clerk of works for assistance in solving site problems. Note, however, that a contractor who acts on the advice or even instructions of the clerk of works, does so at its peril. Giving advice to the contractor is not part of the power or the duty of the clerk of works.

The architect should make sure that the contractor understands the position at the beginning of the contract. It is prudent to give some time to this topic at the first contract meeting and give some space to it in the minutes. On some projects, contractors get into the habit of accepting the clerk of works as speaking on behalf of the architect. There is no justification for this view in the contract and it saves much bad feeling to put the matter beyond doubt. One of the worst things that can happen on a contract is for the architect to have to overrule the clerk of works.

The duties of the clerk of works can, thus, be seen in two parts:

1. Duties under the contract;
2. Duties by virtue of the architect's specific directions.

The clerk of works' duties under the contract define relations with the contractor and the duty to comply with the architect's directions defines relations with the architect. That is to put the matter in broad terms. Every architect has

Box 8.3   Letter from architect to clerk of works setting out duties.

Dear Sir

PROJECT TITLE

My client *[insert name]* has confirmed your appointment as clerk of works for the above contract. I should be pleased if you would call at this office on *[insert date]* at *[insert time]* to be briefed on the project and to collect your copies of drawings, specification, schedules, weekly report forms and site diary.

The contractor is expected to take possession of the site on the *[insert date]*. You will be expected to be present on site for a minimum of *[insert number]* hours per week from that date. I will discuss the timing of your visits when we meet. Let me know at the end of the first week if proper accommodation is not provided for you as described in the specification.

Your duties will be as indicated in the Conditions of Contract special clause *[insert number]*, a copy of which is enclosed for your reference. In particular, I wish to draw your attention to the following:

1. You will be expected to inspect all workmanship and materials to ensure conformity with the contract, i.e. the drawings, specification, schedules and further instructions issued from this office. Any defects must be pointed out to the person in charge, to whom you should address all comments. If any defects are left unremedied for 24 hours or if they are of a major or fundamental nature, you must let me know immediately by telephone.
2. Although I know that it is accepted practice on some projects for clerks of works to mark defective work on site, you must not make such marks or in any way deface materials on site even if you believe they are defective.
3. It is not my policy to issue lists of defects to the contractor before practical completion (commonly known as 'snagging lists'). They are open to misinterpretation and should be compiled by the person in charge. Confine yourself to oral comments.
4. The architect is the only person empowered to issue instructions to the contractor.
5. Any queries, unless of a minor explanatory nature, should be referred to me for a decision. You are not empowered to vary or omit work.
6. The report sheets must be filled in completely and a copy sent to me on Monday of each week. Pay particular attention to listing all visitors to site and commenting on work done and defects observed and notified in as much detail as possible.
7. The diary is provided for you to enter your daily comments.
8. Remember that your weekly reports and site diary may be called in evidence should a dispute arise, so you must bear this in mind when making your entries which should be as full as possible.

The successful completion of the contract depends in large measure on your relationship with the contractor. If you are in any doubt about anything, please let me know.

Yours faithfully

particular views about the relationship with the clerk of works, but among the duties the clerk of works might be expected to fulfil are the following:

- Inspect the works;
- Relay queries and problems back to the architect;
- Complete report sheets;
- Complete daily diary;
- Take measurements as directed;
- Take particular notes of such things as portions of the work opened up for inspection.

3.4

Defacing work

The clerk of works is not entitled to put any marks on defective portions of the work. Some clerks of works make a practice of defacing work, materials or good which they consider to be defective. There are two substantial objections to this practice:

1. Although the clerk of works may consider the item to be defective, the contractor may be able to demonstrate that it is actually fully in accordance with the requirements of the contract.
2. If it is defective, once such work is removed, it is the property of the contractor who is entitled to expect that the clerk of works will not do any damage to the work, no matter how slight.

If this problem does occur, the contractor should write to the architect immediately (Box 8.4).

Box 8.4   Letter from contractor to architect if the clerk of works defaces work or materials.

Dear Sir

PROJECT TITLE

Unfortunately, it is common practice for the clerk of works to deface work or materials considered to be defective. The basis for such action is presumably to bring the defect to the notice of the contractor and ensure that it cannot remain without attention.

We object to the practice because:

1. The work or materials so marked may not be defective and we will be involved in extra work and the employer in extra cost in such circumstances.
2. The work or materials so marked, if indeed defective, will not be paid for and will be our property when removed. We may be able to incorporate it in other projects where a different standard is required. Defacement by the clerk of works would prevent such reuse.

We will take no action about the defacing marks we noted on site today, but if the practice continues, we will seek financial reimbursement on every occasion.

Yours faithfully

Snagging lists

It is, unfortunately, common practice for the clerk of works to issue 'snagging lists' to the contractor, particularly towards the end of a job. There are very rare occasions when there may be pressing reasons for taking this course of action. Generally, the practice should be discouraged because:

- The clerk of works is inspecting for the benefit of the employer, and owes no duty to the contractor to find defects.
- It is the job of the person in charge to produce lists of defective work.
- The contractor may be under the impression, however misguided, that if it simply attends to defects on the 'snagging lists', its obligations are at an end, and disputes may follow.

Obviously, the clerk of works must draw the contractor's attention to work not in accordance with the contract documents, but should not be more specific. In particular, neither the clerk of works nor the architect should instruct how defective work is to be corrected. To do otherwise may result in the employer having to pay for the work.

93

8.3   Responsibilities

Like everyone else connected with the contract, the clerk of works has a responsibility to carry out relevant duties in a competent manner. The clerk of works must demonstrate the same degree of skill that would be demonstrated by the average clerk of works. If a clerk of works purports to be especially skilled in some branch of work (for example plumbing), a greater standard of skill than the average will be expected in that particular branch.

94

If the employer employs and pays the clerk of works directly, the employer will be liable for the actions of the clerk of works in the normal way. The fact that the clerk of works is under the architect's direction makes no difference. The relationship between the clerk of works and architect has been compared to that between the chief petty officer and the captain of a ship. This does not mean that, if the clerk of works is negligent, it will relieve the architect of all responsibility, but it may substantially reduce the architect's liability for damages depending on circumstances. It is, therefore, very important to ensure that the employer employs the clerk of works. The negligence of a clerk of works employed by the architect will not reduce the liability of the architect.

8.4   Common problems

The clerk of works has not notified the contractor of defects in the Works, they are not defects, merely snags which can be picked up later.

There is some kind of myth around that a snag is a lesser kind of defect. The contract makes no reference to 'snags' or 'snagging lists'. The contract does refer to defects. A defect is when something is built which is not in accordance with the contract documents. There are minor defects and major defects, but there are no snags. Contractors often refer to 'a few snags' when they intend to play down any irregularities. The truth is of course that a defect or a snag is a breach of contract on the part of the contractor and if the contractor will not rectify it, the employer is entitled to have it corrected by another contractor and recover the cost. Whatever label it is given, a defect is a defect and a breach of contract.

2.10 (2.11), 3.5

The clerk of works tells the contractor that he has put a 'stop notice' on the work.

This does not appear to be quite as common as it once was. The clerk of works has no authority to instruct the contractor to stop work or to do, or to desist from, doing anything. A contractor which stopped work because the clerk of works gave a 'stop notice', would be in breach of contract.

On some projects it may be very difficult for the contractor to ignore the clerk of works, especially if the architect regularly uses the clerk of works to pass on instructions. The only way is for the contractor to explain the position and inform the architect very clearly that instruction will only be accepted and carried out if given personally by the architect.

9 Sub-contractors and suppliers

9.1 General

3.1, 3.3 The contract provisions are extremely brief. There are no provisions for suppliers, employer's licensees or nominated or named sub-contractors. The possibilities of having persons engaged directly by the employer or nominating or naming sub-contractors are considered below.

9.2 Differences between assignment and sub-contracting

Assignment is usually coupled with sub-contracting in contract provisions and MW and MWD are no exception. It is a mistake, however, to consider that they are linked. They are totally different concepts. Assignment is the legal transfer of a right or duty from one party to another as a result of which the original party retains no interest in the right or duty. For this to be fully effective a new contract must be formed (i.e. *novation*).

Sub-contracting is, in essence, the delegation of a duty from one party to another, but the original party still retains primary responsibility for the discharge of that duty. It is vicarious performance of a duty by someone else.

Suppose that an architect had the duty to design foundations. It could be agreed with the architect's client that the architect would assign the duty to a structural engineer. From that time onward, the architect would not be responsible for the design of the foundations, it would be the engineer who would be responsible. However, if it was simply agreed that the architect would sub-contract the design to the engineer, the architect would still be responsible for the design so far as the client was concerned. The engineer would be responsible to the architect.

9.3 Assignment

3.1 Both employer and contractor are prohibited from assigning the contract unless one has the written consent of the other. This is a much stricter provision than

The JCT Minor Works Building Contracts 2016, Fifth Edition. David Chappell.
© 2018 John Wiley & Sons Ltd. Published 2018 by John Wiley & Sons Ltd.

is to be found under the general law where it is possible for either party to assign the benefits or rights of a contract to a third party. For example, the contractor might wish to assign to a third party the benefit of receiving progress payments in return for substantial financial help at the beginning of the contract. The employer, too, may wish to sell an interest in the completed building to another before the issue of the final certificate, thus assigning the benefit of the contract. Although assignment of rights or benefits is permitted under the general law, the contract is worded to stop such deals. Duties or obligations of contracts can never be assigned without express agreement between the parties. This kind of provision is effective to prevent assignment of the benefits of a contract or the right to damages for breach of the contract.

It is considered that the architect has a duty to explain the possible difficulties to the employer, particularly if there is reason to believe that the employer might wish to assign a benefit before the final certificate is issued. Assignment can be made with the written consent of the other party, but such consent may be refused and the grounds for refusal need not be reasonable. It would appear to do no harm, and make much sense, for the architect to advise the employer to have the contract amended so as to prohibit only the assignment of duties under the contract. The amendment should be carried out at tender stage and, if anything, should result in lower rather than higher tenders.

9.4 Sub-contracting

Sub-contracting is traditional in the building industry, but the practice can be abused to the extent that the contractor's sole employee on the site might be the person-in-charge, the remainder of the workforce being sub-contractors. Needless to say, such an arrangement does not make for an efficient contract. The contract provisions are designed to prevent this and other problem situations by allowing sub-contracting only if the architect gives written consent (see Box 9.1). The architect may withhold consent but

---

Box 9.1   Letter from contractor to architect requesting consent to subletting.

Dear Sir

PROJECT TITLE

We propose to sub-let portions of the Works as indicated below, because [*state reasons*]. We should be pleased to receive your consent in accordance with clause 3.3.1.

[*List the portions of the Works and the names of the proposed sub-contractors*]

Yours faithfully

not unreasonably. It is probably reasonable for the architect to require the contractor to supply the name of the proposed sub-contractor before consent is given.

It is important to remember that there is no contractual relationship between the employer and the sub-contractor. The sub-contractor's contract is with the contractor. It follows that nothing contained in MW or MWD is binding in any way on the sub-contractor. It is vital, therefore, that the sub-contract gives the contractor sufficient controls over the sub-contractor because the employer must look to the contractor for redress if the sub-contractor defaults.

There is now a standard form of Minor Works Sub-Contract with sub-contractor's design for use with MWD. The JCT Short Form of Sub-Contract is suitable for use with MW although the contract simply says that the relevant sub-contract should be used 'Where considered appropriate'. Therefore, the contractor is not obliged to use either sub-contract. Nevertheless, the contractor must include three specific terms in any sub-contract even if it is a contractor's 'home grown' sub-contract. The terms are that:

- The employment of the sub-contractor must terminate immediately the contractor's employment under MW or MWD terminates for any reason,
- Each party must comply with the applicable CDM Regulations,
- If the contractor fails to pay an amount due to the sub-contractor by the final date for payment, the contractor will pay the sub-contractor simple interest at 5% over Bank of England official bank rate.

It is thought that the architect would not be unreasonable to withhold consent to sub-letting until satisfied that the contractor's sub-contract provisions were adequate. The architect should beware, however, of being drawn into disputes between contractor and sub-contractor or of being seen to approve the form of sub-contract. It should be remembered that ultimately the contractor is responsible for its own sub-contractors and their performance. It has no right to look to the architect or the employer to assist if things go wrong. The contract contains a term which confirms that, despite any sub-contracting, the contractor is wholly responsible for carrying out and completing the Works. In light of the meaning of sub-contracting, it is questionable whether this term is strictly necessary.

The lack of detailed provisions is sometimes to be regretted, but is unavoidable taking the contract as a whole. Unfortunately, although the number of sub-contractors will tend to increase with the size of the work, there is no lower point at which the contractor will not use any sub-contractors at all.

9.5 Nominated sub-contractors

MW and MWD make no provision for nominated sub-contractors. It is not envisaged that there will be any need or requirement to nominate sub-contractors for work for which this form would be suitable. Having said that, there are various devices which can be used to provide for 'nominated' sub-contractors if the architect or the employer is absolutely sure that they must 'nominate' a particular firm.

The widespread use of nominations in building contracts has come to be associated with certain laziness or shortage of time on the part of the architect. It often seems quicker and less trouble in the short term to put a sum in the bills of quantities or specification than to properly specify requirements at the outset. It is obviously a temporary expedient only and at some future date, usually sooner rather than later, the architect is faced with specifying the work in question and engaging in the complicated process of nomination. Difficult legal and administrative problems can result during the life of the contract – as numerous legal cases testify. The message is clear. Wherever possible, nomination should be avoided. If nomination is required consideration should be given to using IC or ICD, which provide for 'naming' sub-contractors in a clause of complexity and some obscurity, or ACA 3, which has relatively simple 'naming' procedures. If 'nomination' or something similar is desired using MW or MWD, it is possible to use one of the following methods:

- Name one or a choice from several named firms in the specification;
- Name a firm in an instruction directing the expenditure of a provisional sum;
- Include an appropriately worded clause in the contract;
- Provide for the specialist firm to be directly employed by the employer.

3.7

There are severe pitfalls inherent in the use of any of these methods and the architect should seek the advice of an experienced construction contract consultant before proceeding. Among the points to be considered are the following:

Naming in the specification

It is perfectly possible for the architect to name one firm in the specification which the contractor must use for carrying out a specific part of the work. If this is done, a considerable amount of power is placed in the hands of such a firm which can, in effect, quote to the contractor whatever price it likes, secure in the knowledge what whichever contractor is awarded the contract, the specialist firm will be incorporated. If, as is not unknown, the specialist firm goes into liquidation before work begins on site, the architect has an instant problem with which to start contract administration duties. In such circumstances, the architect will have to move quickly to issue an instruction varying the name of the sub-contractor, perhaps doing so more than once to deal with possible objections from the contractor. The answer, of course, is to incorporate a fairly substantial clause in the contract to deal with the situation. Such a clause would require very careful drafting.

The alternative, whereby the contractor is given a choice of three or four names in the specification, is not strictly nomination at all. However, it does give a degree of control over the firms to be used, while allowing the contractor to seek competitive quotations. A sub-contractor appointed by the contractor under this method would be the contractor's entire responsibility to the extent that, if it failed, it would be the contractor's job to find an alternative. Although the worst consequences of nomination can be avoided by this system, it is advisable to include a suitably worded clause in the contract.

Naming by an instruction in respect of a provisional sum

If it is intended to operate this system, it is essential that the contractor knows what is intended at the time of tender so that it can make suitable provision in its price. A big danger is that the contractor might strongly object to the firm in question when it is named for the first time in an instruction. There is no machinery to deal with its objection and the efficient progress of the work can be seriously affected. All the comments regarding the naming of a single firm in the specification are equally applicable to this case.

Including a clause in the contract

The trouble with this is that once substantial clauses are added dealing with particular circumstances, in this case nomination, there is the risk of the contract being considered the employer's 'standard written terms of business,' which brings it within the scope of the Unfair Contract Terms Act 1977. The clauses are also liable to be interpreted against the employer if there is any ambiguity. There is no doubt that a suitable clause can be formulated and, if nomination is what is wanted, it is probably the safest way to achieve it. Nomination can proceed along time-honoured procedures and potential difficulties can be provided for. On the other hand, to add an effective nomination clause would effectively double the size of the contract – to say nothing of the ancillary documentation which would be necessary.

The employer directly employing the specialist firm

This may seem an attractive solution, but there are dangers, not least the problem of integration with the Works (see Chapter 10).

The best way is not to try to nominate or name using this form of contract.

9.6   Common problems

Under MW the architect has specified a sub-contractor to design, supply and fix the staircase. The sub-contractor has fitted a staircase, but it does not comply with the building regulations.

It is always dangerous to specify a sub-contractor by name, as noted above. The architect must now give instruction to the contractor to have the staircase removed and re-designed. The problem here is that, because the architect has used MW instead of MWD, the contractor has no design responsibility for the staircase. The sub-contractor is responsible to the contractor for the design, but the contractor is not responsible to anyone for the design. Unless the architect

96

has previously agreed with the employer that the sub-contractor should do the design, the architect will be responsible to the employer for it.

No doubt the contractor will do its best to get the sub-contractor to alter the staircase, but if the contractor has notified the architect of the divergence from statutory requirements, it will not be liable for the divergence under the contract. If the sub-contractor chooses this moment to go into liquidation, the architect will have a big problem. The staircase will have to be removed and redesigned to comply with building regulations and the employer will have to pay. The architect will have to do the redesign and most likely the employer would expect the architect to pay for the rebuilding.

2.5.2

If the employer finds a firm which will do the electrical work which is in the contract at a price much cheaper than the contractor would do it, does the architect have power to omit the work from the contract?

97

The architect certainly does have that power, but it is a power to omit work, not to transfer work to another contractor without good reason. Finding a cheaper electrician after the contractor's price has been accepted is not a good reason. As soon as the employer engaged the other electrician, the employer would be in breach of contract. The contractor would be entitled to damages to reflect what it had lost (usually the profit element).

There would be an added complication, because strictly the electrician would not be entitled to go onto the site without the agreement of the contractor. Even if the contractor made no objection, it would have no responsibility for the electrician nor for organizing when the electrician did work. If the electrician delayed the contractor by not turning up when electrical work was required, the contractor would be entitled to an extension of time and, if it cost the contractor money, the cost would be part of its damages for breach of contract.

10 Statutory matters and work outside the contract

10.1 Statutory authorities

2.1.1 (2.1)

The contract states that the contractor must comply with statutory requirements, which are defined as including any statute, any statutory instrument, rule or order or any regulation or bye-law applicable to the Works. This means, for example, that the contractor must not contravene the Planning Acts or regulations made under those Acts, and must comply with the Building Regulations and, of course, the Construction (Design and Management) Regulations 2015. The contractor must also comply with any regulation or bye-law of a local authority or statutory undertaker having jurisdiction for the Works such as electricity, gas and water suppliers. By necessary implication, therefore, the contractor must allow statutory undertakers to enter the site and carry out work which they alone are empowered to do.

Certain crucial parts of virtually all contracts are carried out by statutory undertakers such as local authorities, gas, water and electricity suppliers. When they carry out work solely as a result of their statutory rights or obligations, they are in a special category quite separate from sub-contractors or firms directly employed by the employer. If completion of the Works is delayed as a result of a supplier carrying out or failing to carry out work which is part of its

2.7 (2.8)

statutory duty, the contractor will be entitled to an extension of time.

If the contractor employs a statutory undertaker to carry out work which is not part of its statutory rights or duties, the statutory undertaker ranks as an ordinary sub-contractor. Delay caused by the undertaker in those circumstances would not entitle the contractor to an extension of time under MW or MWD. An example should make the principle clear.

It is usual to include a provisional sum in the specification to cover the cost of connecting the electrical system of a dwelling to the mains. The contractor may also, with the architect's written permission, sub-let the electrical wiring and fittings in the dwelling to the electricity supplier. The mains connection is part of the supplier's statutory obligations; the internal wiring and fittings are not part of the supplier's statutory obligations, but a matter of contract between the contractor and the supplier. If the mains connection delays the completion date, the contractor is entitled to an extension of time. If the internal wiring and fittings delay the completion date, the contractor is not entitled to any extension of time.

Statutory undertakers have no contractual liability when they are carrying out their statutory duties – a sore point with many architects – but in certain cases they have a liability outside the contract in tort. Outside their statutory duties, they are in the same position as anyone else if they enter into a contract to carry out work.

Notices and fees

2.1.1, 2.6
(2.1, 2.7)

3.6

The contractor must also give all notices required by statute etc. and must pay all fees and charges in respect of the Works, provided that they are legally recoverable from the contractor, but not otherwise. The contractor is not entitled to be reimbursed and is deemed to have included the necessary amounts in the price. The exception to this, of course, is if the charge is a necessary result of an architect's instruction. In such a case, the amount of the charge will form part of the valuation of the instruction.

Liability

2.5.2 (2.6.2)

The contractor is not liable to the employer under the contract if the Works do not comply with statutory requirements provided that:

- The contractor has carried out the Works in accordance with the contract documents or any of the architect's instructions; *and*
- If the contractor has found a divergence between the contract documents or the architect's instructions and statutory requirements, the contractor has immediately given him a written notice specifying the divergence.

98

The contractor is not liable if it fails to find a divergence which actually exists. Its duty is not to look for divergences but simply to report them if it becomes aware of them. Although the contractor may be freed from liability to the employer, its duty to comply with statutory requirements remains. Thus, the local authority may serve notice on the contractor if work, built correctly in accordance with the contract documents, does not comply with the Building Regulations. In such a case, the architect would have to act speedily to issue appropriate instructions.

Emergencies

The contractor is not entitled to take any emergency measures to comply with statutory requirements even though delay might cost the employer money. The contractor's obligation to give the architect immediate written notification remains. If the emergency concerns part of the structure which is actually dangerous, the contractor has a responsibility under the general law to take whatever measures are necessary to make the structure safe. It is difficult to see how the contractor could make any valid claim in such circumstances.

CDM Regulations

3.9

The contract makes provision for compliance of the parties with the CDM Regulations. The idea is to make compliance with the regulations a contractual duty so that breach of the regulations is also a breach of contract. There is a provision that the employer 'shall ensure' that the principal designer carries out all the relevant duties under the regulations and that, where the principal contractor is not the contractor, the employer must ensure that it will also carry out

3.9.1

its duties in accordance with the regulations. There are also provisions that the contractor, if it is the principal contractor, will comply with the regulations.

Every architect's instruction potentially carries a health and safety implication which should be addressed. The CDM Regulations pose a formidable list of duties on the parties. Some of these duties must be carried out before work is started on site. If necessary actions delay the issue of an architect's instruction or once issued delay its execution, the contractor will be entitled to extension of time and, depending on circumstances, it may have a common law claim for damages for breach of an express term of the contract. These are matters about which all parties should take great care.

10.2   Works not forming part of the contract

The contract makes no provision for the employer to enter into a contract with anyone other than the contractor to carry out any part of the work on site while the contract Works are being carried out. It is quite common for the employer to wish to engage others to do certain work or, indeed, to use some of its own employees. The reason may be because the employer has a special relationship with the firm or individual, for example in the case of a sculptor, artist or landscaper, or because the employer wants complete control over a particular operation. MW and MWD do not allow for direct engagement of another contractor by the employer, but it can be achieved in one of two ways:

- With the consent of the contractor; or
- By including a special clause to that effect.

It is never a good idea to bring third parties onto a site during the contract period. Three dangers which merit special consideration are:

1. It is easy for the contractor to claim that such persons have disrupted its work and/or delayed the completion date. They are very difficult claims to refute because introducing third parties clearly does not help the contractor and can usually be seen to cause at least some degree of hindrance. The contractor may be able to claim damages at common law and an extension

2.7 (2.8)

   of time.

2. The contractor may acquire grounds to terminate its employment under the contract. The contract provides for the contractor to terminate if the Works are suspended for a continuous period of one month because of any impediment, prevention or default of the employer, the architect or of

6.8.2.2   any employer's persons. The employment of other contractors would certainly fall into this category. Whether the Works would be delayed by as much as a month would depend on the circumstances. This matter is dealt with in more detail in Chapter 21.

1.1   3. 'Employer's Persons' are persons for whom the employer is responsible. They are not sub-contractors. Therefore, the employer may have uninsured liabilities. The directly employed persons may well have their own insurance cover, but it is the architect's duty to advise the employer to obtain the necessary cover through the employer's own broker. The cover should be for the employer and those persons for whom the employer is responsible in respect of acts or defaults occurring during the course of the work. It is a complex business and best left in the broker's hands. It is not the architect's responsibility to advise on insurance matters.

Statutory undertakers acting outside the confines of their statutory duties may be considered to be the responsibility of the employer if they are not paid by the contractor and under its control. It is always prudent to ensure that the contractor is responsible for all the work to be carried out during the currency of a contract. It removes some possible areas of dispute and promotes efficiency. If the employer insists on having directly employed persons, remember that, for work to be considered as not forming part of the contract, it must:

- Be the subject of a separate contract between the employer and the person who is to provide the work; *and*
- Be paid for by the employer direct to the person employed, not through the contractor.

10.3   Common problems

The specification states that the employer will supply the roofing slates.

There is nothing in the contract to allow the employer to supply materials. The contract quite simply assumes that the architect will supply the specification and drawings and the contractor will provide the necessary labour, materials and equipment to construct the building designed by the architect. Where something has been reserved for the contractor to design, it and not the architect will be responsible for designing that part. However, if the contractor has priced on the basis that certain materials, in this case slates, will be supplied by the employer, both parties are bound by that agreement. If the employer provides the slates when required by the contractor and if the slates are of a satisfactory quality, there should be no difficulties.

However, life being what it is, the chances are that the employer will not be able to deliver the slates on time and and/or they may not be of a satisfactory standard. That is when the problems start. If the contractor was responsible for supplying all materials, the late delivery and poor quality of the slates would be the contractor's problem. In this case, timing and quality are outside the contractor's control and, therefore, entitle the contractor to an extension of time if the date for completion is delayed. There is no contractual mechanism to allow the contractor to claim loss and/or expense for the delay or disruption to the Works, but the contractor would certainly have a claim for breach of contract at common law against the employer. The employer's idea for saving money might eventually cost more than if the ordinary contract procedures, tested over time, had been used.

The building control officer (BCO) has been to site and instructed the contractor that the foundations to a house must be excavated a further one metre deep before concrete is poured.

99

This scenario is quite common and can lead to problems. The position is clear that if the contractor receives an instruction from the BCO, the contractor must inform the architect who must decide what to do about it. Many contractors think that in these circumstances they have no choice but to follow the instructions of the BCO, who, after all is responsible for overseeing the application of the Building Regulations. The reality is that there may be several ways of satisfying the Building Regulations of which excavating a further one metre is but one way. The architect may decide on a completely different foundation system, such as a raft or short piles as a better and cheaper solution. If the contractor simply carries out the instruction of the BCO, it will be in breach of contract and, if the architect allows the extra concrete fill to remain, the contractor will not be paid, unless that is the only solution.

11 Insurance

11.1 Important

Insurance is very complicated, with its own rules. The architect is not usually an expert in insurance matters. The expert should be the employer's insurance broker. Although the architect must be able to explain the purpose of the insurance clauses to the employer, other questions on insurance must be referred by the employer to the broker. The architect is expected to be able to point to the standard clauses, but often there are problems in insuring work to existing property which are not covered by the standard clauses. The architect should beware of giving any opinion about insurance matters and should always refer the employer to the insurance broker or another expert. A problem which frequently arises is when the contractor is required to provide details of its insurance. The architect must not fall into the trap of attempting to interpret the wording of the contractor's policy; neither is it sufficient simply to pass the policy to the employer (often referred to as acting like a postbox). Either of these courses of action may result in the architect becoming liable for the giving of bad advice to the employer. The architect should pass the insurance documents to the employer with a note to the effect that the architect is not an insurance expert and suggests that the employer seeks the advice of a broker on the adequacy of the policy. The architect could seek the advice, but that is not to be recommended. It is better if the architect keeps such matters at arm's length.

It is always wise for the employer to retain the services of an insurance broker to give advice about the insurance provisions of the contract. The broker inspects all the relevant documents and certifies to the architect that they comply with the contract requirements. It is essential that the insurance is in operation from the time the contractor takes possession of the site.

There are two important definitions: 'Contractor's Persons' and 'Employer's Persons:'

- In the case of the contractor, they include employees, agents, sub-contractors and anyone on site at the contractor's behest, but not including the architect, employer, employer's persons or statutory undertakers.
- In the case of the employer, they include anyone authorised by the employer, but not the architect or statutory undertakers.

The JCT Minor Works Building Contracts 2016, Fifth Edition. David Chappell.
© 2018 John Wiley & Sons Ltd. Published 2018 by John Wiley & Sons Ltd.

11.2 Injury to or death of persons

The contract states that the contractor is liable for any expense, liability, loss, claim or proceedings in respect of personal injury or death of any person resulting from carrying out the Works, except to the extent it is due to act or neglect of the employer, any employer's person or any statutory undertaker. Note that employer's person does not include the architect. Moreover, the contractor must indemnify the employer against that kind of expense, liability etc. To say that the contractor must indemnify the employer against something means that the contractor must make good the employer's losses.

In practical terms, the employer will be responsible for the injury or death of any person only insofar as the injury or death was caused by the employer or the employer's persons' act or neglect. The contractor retains liability even if the employer is to some degree responsible. In effect, in such circumstances, the employer makes a contribution to reflect the extent of his or her negligence. Therefore, if a pedestrian walking along the public footpath outside the building suffers injury due to debris thrown out of a window by one of the contractor's employees, that pedestrian might well sue the employer for medical costs, loss of earnings and so on. The employer would then join the contractor as a third party in any action and claim an indemnity under this clause. If the contractor could show that the employer had contributed to the injury by some negligent action (unlikely in this example) the contractor's payment to the employer would be reduced accordingly.

Insurance

The contractor must take out and maintain and cause any sub-contractor to take out and maintain insurances which must comply with all relevant legislation in connection with claims for personal injury or death of any person employed by the contractor and which arise from and during the person's employment and for all other claims for personal injury or death for which the contractor indemnifies the employer. There is a space in the contract particulars for the insertion of a suitable sum. The opinion of an insurance broker should be sought by the employer as to the amount which should be included. The contract gives the employer the right to require evidence that the insurance has been taken out and maintained. The employer must give seven days' notice of that requirement to the contractor. There is no provision for the employer to take out the necessary insurance and deduct the amount from the Contract Sum if the contractor defaults. Nonetheless, failure on the part of the contractor to insure is a breach of contract for which the employer could recover damages at common law, provided of course the damage suffered could be properly evidenced. In practice, it is certain that the employer would set off such sums against money payable to the contractor after first issuing a pay less notice (see Chapter 18).

It usually falls to the architect on behalf of the employer to ask the contractor for evidence of insurance. It is commonly thought that if the contractor fails to take out insurance, it has no liability to the employer. That notion is entirely wrong. The requirement that the contractor insures is stated to be without affecting its liability to indemnify the employer. In the case of an insurance company failing to pay in the event of an accident, the contractor is bound to find the money itself. The real problem is that, in the absence of insurance, the contractor may be unable to find the money. The requirements of the contract should be satisfied if the contractor and sub-contractors already have general insurance cover in appropriate terms in an adequate amount.

11.3   Damage to property

The contractor assumes liability for, and indemnifies the employer against any expense, liability, loss, claim or proceedings arising out of the carrying out of the Works in respect of injury or damage to any kind of property other than the Works, any unfixed materials and goods on the Works to the extent that it is due to the negligence, breach of statutory duty, omission or default of the contractor or any contractor's persons engaged by the contractor upon the Works.

Insurance

If the employer is taking out joint names insurance for the Works and existing structures, the contractor is not liable nor need he indemnify in the case of damage to existing structures and contents caused by specified perils. That is the case even if the loss or damage is due wholly or partly to the negligence of the contractor or a contractor's person.

The contractor's liability under this clause is limited compared with its liability for personal injury or death. The contractor or sub-contractor must be at fault for the indemnity to operate. The contractor need not be totally at fault for indemnity purposes. If it is partly at fault, the employer has a partial indemnity. The contractor must insure and cause any sub-contractor to insure against his liabilities.

If the Works and existing structures are insured by other means, any limitations and exclusions in the insurance will govern the contractor's liability and indemnity in respect of loss or damage to the existing structure and contents.

In general, the position appears to be that if the existing building and contents are the subject of damage by specified perils, they should be covered by the insurance even if wholly due to the contractor's negligence. Damage caused by matters other than specified perils must be rectified at the expense of the contractor if due to its negligence, otherwise it is at the risk of the employer. A fire caused by the contractor's negligence is not covered by the employer's insurance and must be dealt with under the contractor's own insurance.

The contract provides for the employer to stipulate the amount of cover required. The general comments with regard to inspection of documents in section 11.1 are also applicable to this clause.

Non-negligent damage to other property

There is no provision in MW or MWD for insurance against claims arising due to damage to other property caused by the carrying out of the Works when there is no negligence or default by any party. Most standard forms of contract contain such provision which can be invoked if required, to cover specific risks, for example the carrying out of underpinning works to adjoining property. It is not always necessary to take out such insurance and the provision is probably omitted from the contract in view of the minor nature of the works envisaged. The amount of the Contract Sum, however, is no indication of the possible risk to neighbouring premises and the architect would be prudent to assess each project on its own merits. There is no reason why the insurance should not, and every reason why it should, be taken out by the employer with the advice of a broker if circumstances appear to warrant it.

103

11.4   Insurance of the Works

Important terms

Although the architect and contractor should be aware of these definitions and what they mean in general terms, detailed advice on them is a matter for the employer's insurance broker.

Joint names

A joint names insurance policy is one in which two or more people are named as insured. In the case of MW and MWD, it is a policy in which the employer and contractor are named as insured. The key thing is that the insurers have no rights against either employer or contractor. This is very important. The usual position is that, if a claim is made by the insured person against the insurance company, the insurance company after paying out may stand in the shoes of the insured person (called 'subrogation') and claim against whoever was actually responsible for the loss. If the insurance is in joint names, the insurer cannot claim against the contractor even if the contractor is responsible for the damage.

Excepted risks

The contractor is not liable to indemnify the employer or to insure against damage to the Works, the site, or any property against damage due to nuclear perils and the like.

1.1

104

All risks insurance

The key point about 'all risks insurance' is that it must provide cover against 'any physical loss or damage to work executed and Site Materials'. Cover is not limited to 'specified perils' (see below). It includes other risks such as impact, subsidence, theft and vandalism.

Specified perils insurance

'Specified perils' are fire, lightning, explosion, storm, flood, escape of water from any water tank, apparatus or pipe, earthquake, aircraft and other aerial devices or articles dropped from them, riot and civil commotion, excluding any loss or damage caused by the excepted risks.

Alternative insurances

Clause 5.4 deals with the insurance of the Works against a collection of specific risks. The clause is divided into three parts, only one of which is to apply. Each part is applicable to a particular situation as follows:

- 5.4A: Joint names insurance by the contractor for new Works.
- 5.4B: Joint names insurance by the employer for Works and existing structures.
- 5.4C: Insurance of the Works and existing structure by some other means.

The contract particulars must be completed to indicate which part is to apply.

A new building where the contractor is required to insure in joint names

5.4A   The provision is not complex and provides for the contractor to insure in the joint names of the employer and itself against loss or damage caused by all risks. The insurance must cover the full reinstatement value and include:

- All executed Works; and
- A percentage to take account of all professional fees (the percentage must be stated in the contract particulars and will be 15% unless another figure is stated).

and exclude temporary buildings, plant, tools and equipment owned or hired by the contractor or his sub-contractors. (This is clear although the contract does not expressly refer to it.) These items are not part of the Works and should be covered under the contractor's general insurance arrangements.

The insurance must be kept in force until the date of issue of the certificate of practical completion or termination of the contractor's employment. This is the case even if practical completion is delayed beyond the date for completion in the contract. If the contractor already maintains an 'All Risks' policy which provides the same type and degree of cover required, it will satisfy the contractor's obligation to insure if it is noted as recognising the employer as joint insured. It should be noted that the contractor's 'All Risks' policy will not necessarily insure against all the employer's perceived losses, such as loss of revenue. Depending on circumstances, the contractor may not be liable for

105   such losses in any event.

Alterations or extensions to existing structures

5.4B   In the case of work to an existing building, the employer must insure the Works against all risks in joint names. The existing structures and contents and all unfixed materials must be insured only against specified perils. The contractor's temporary buildings etc. must be excluded although the contract does not mention them. The employer need only insure contents that are owned by him or her, or for which the employer is responsible. The reason for this is not clear. It would seem sensible that the employer's insurance of contents includes things which are on the premises but which do not belong to the employer. The employer must insure for the value of professional fees. The percentage must be stated in the contract particulars and will be 15% unless another figure is stated. The employer's obligation to insure in joint names ceases on practical completion and the contractor is open to claims for damages from the employer.

106

Insurance of the Works and existing structure by some other means

5.4C

5.4B

This provision acknowledges that it will not always be possible for insurance to be obtained by the employer to comply with the standard set out in the contract. The idea is that, before tenders are invited, the employer discusses the situation with his or her insurance broker who can advise on the best arrangement to safeguard the interest of both employer and contractor. The details can then be entered in the contract particulars. If, as seems likely the details are too lengthy to put in the contract particulars, details of an attachment to the contract can be entered there along with the reference to clause 5.4C. There are many possible complications involving the employer as landlord or as tenant in a block of flats or a company tenanting a whole or part of a building and so on.

11.5   Evidence of insurance

If the employer or the contractor is required to take out insurance or make sure that it is taken out, the other party may request evidence of the insurance. The one responsible for insuring must respond within seven days providing written evidence that the insurance has been taken out and/or is being maintained. The party wanting the evidence is entitled to request whatever is reasonable. What is reasonable will depend on the circumstances and, once again, the architect, if asking for the evidence, must first establish from the employer's broker what such reasonable evidence might be.

The architect must not simply pass the documents to the employer without comment. As noted earlier, the architect must do one of three things:

- Give advice to the employer, something which the architect is unlikely to be qualified to do; or
- Obtain expert advice on the policy and pass this on to the employer; or
- Advise the employer to obtain expert advice.

11.6   Loss or damage

107
5.6.1
If damage occurs, the contractor must notify the architect and the employer as soon as it reasonably can do so. If the damage affects executed work or materials on site which the contractor is responsible for insuring, the contractor must 'with due diligence' restore or replace work or materials or goods damaged, and dispose of debris before proceeding to carry out and complete the Works as before. If the insurers wish to carry out an inspection, the contractor must wait
5.6.4
until the inspection is finished before proceeding.

The contractor must authorise the insurers to pay the employer all monies from the Works insurance policy and any other policies covering existing struc-
5.6.3
tures and contents whether taken out by the contractor or the employer.

If the contractor has taken out the insurance policy for the Works, the contractor is entitled to be paid all the money received from insurance less only
5.6.5.2
the percentage to cover professional fees. The money is to be released in instalvments on special reinstatement certificates, issued at the same time as interim
5.6.5.1
certificates, but retention must not be deducted. The contractor is not entitled to any other money in respect of the reinstatement and if there is any element
5.6.5.3
of underinsurance or excess, it must bear the difference itself. Interim certificates must continue to be issued in the usual way and the architect must ignore that fact that some of the work carried out by the contractor has since been
5.6.5.2
damaged. In other words, if the contractor has erected the roof carcass but it has been totally destroyed, the contractor must still be paid for the roof carcass as though it was work properly executed.

If the employer has taken out the insurance policy for the Works or if the loss or damage is caused by one of the excepted risks, the reinstatement must be
3.6
treated as though it was a variation. Therefore, the contractor will be entitled to
5.6.6
proper payment for it.

The employer is not obliged to reinstate the existing structure which has
5.7
suffered loss or damage.

Termination

There is provision for either party to terminate the contractor's employment if there is significant damage to existing structures. The contract says that either party may terminate within 28 days of the loss or damage occurring if it is just
5.7
and equitable to do so. The notice must be sent by special or recorded signed-for delivery or delivery by hand so that there is no doubt that the contractor has received it. What is just and equitable will depend on the circumstances. It seems that this provision envisages a situation in which a very serious amount of damage has been suffered which makes the continuation of the project impracticable. Whoever receives the termination notice has seven days in which to set in motion one of the dispute resolution procedures in the contract, otherwise the termination will be deemed to be just and equitable
5.7.1
(the consequences are set out in Chapter 21).

11.7   Common problems

The Works involve an extension and refurbishment to an existing house for the owners and the owners cannot get their insurers to provide insurance under clause 5.4B. However, the contractor says that its insurance will cover everything and has sent copies of the insurance policies to the architect. The employers are happy with this and they are ready to sign the building contract.

Few architects are experts about insurance. If the employers are anxious to proceed and the contractor is happy to be responsible for all the insurance, what is the problem? The problem is that what is being proposed does not fit neatly into the options set out in the contract for insurance of entirely new Works by the contractor or insurance of new Works and existing property by the employers.

5.4A

5.4B   Fortunately, the new edition of MW and MWD allows for the parties to the contract to make their own arrangements and insert the details in the contract

5.4C   particulars. It follows that, if the parties do make their own arrangements, those arrangements must be sufficient to cover the risks involved and the contract entry in the contract particulars must be effective to place the necessary duties on each party.

Architects in this situation must immediately pass the proposal from the contractor to the employer and request the employer to seek expert insurance advice on the situation, probably from the employer's broker. The employer must also be requested to provide the necessary wording for the contract particulars entry, either through the broker or through a solicitor who is either versed in insurance matters or in consultation with the broker. Architects must also point out to the employer that the contract cannot be signed until these matters have been satisfactorily resolved and the contractor cannot commence activities on site until the contract is signed. This is a case where a letter of intent is not appropriate.

Evidence of insurance must be provided by one party to the other within seven days of a request. Then what?

If the evidence is provided, the architect should pass it to the employer with a request that the employer has it verified by the employer's insurance expert. Where the contractor has requested it, it should be vetted by the contractor's broker. If the evidence is not provided, the contract does not say what is to happen. It is not one of the grounds for termination and the sensible thing to do, after due warning, is for the other party to take out the relevant insurance and charge the cost to the defaulting party. The employer can simply deduct it from a payment after sending a pay less notice and the contractor can simply submit it as part of its application for payment.

12 Possession of the site

12.1 Important points

If there is no express term in the building contract, a term will always be implied that the contractor must have possession in sufficient time to allow it to finish the Works by the contract completion date. To have possession of something is the next best thing to ownership. If the owner of this book lends it to a friend, the friend can defend his or her claim to it against anyone except the original owner. The same principle applies to a contractor in exclusive possession of a site. It is in control of the site and has the power to refuse access to anyone else, including the employer. In practice, this stern rule is modified by numerous statutory regulations, allowing entry by the representatives of various statutory bodies, and also by the express and implied terms of the contract.

Licence

In legal terms, the contractor is said to have a licence from the owner of the site to occupy it for the period of time necessary to carry out and complete the Works. The period of time is the period stated in the contract or any extended period. During the contractor's lawful occupation of the site the employer has no power under the general law to revoke the licence, but the contract may contain terms to deal with various situations. For example, in the case of lawful termination of the contractor's employment by either the employer or the contractor itself. If the contract did not say that the employer could retake possession, awkward problems might conceivably arise if the contractor refused to give up possession of the site. In general, however, if the contractor retains possession of the site after the contract period or any extended period has expired, it is in the position of a trespasser and can be removed under the general law.

Retaking possession

This contract contains a term which allows the employer and any replacement contractor to take possession of the site after the employer has terminated.

The JCT Minor Works Building Contracts 2016, Fifth Edition. David Chappell.
© 2018 John Wiley & Sons Ltd. Published 2018 by John Wiley & Sons Ltd.

6.8, 6.9   Although there is no similar specific requirement when the contractor terminates its own employment, a term to that effect will be implied.

There is no provision to allow the architect access to the Works, but such a term would be implied to allow the architect and authorised representatives to carry out their duties under the contract.

12.2   Date for possession

The contract does not make reference to the contractor taking possession of the site. However, the contract particulars do leave a space for a date to be inserted for the rather awkwardly termed 'Works commencement date'. This must be the latest date on which the employer must give possession of the site to the contractor. The date to be inserted in the contract particulars is the earliest on which the contractor will be allowed to commence carrying out the Works, but strangely there is nothing in the contract which requires the contractor to com-

2.2 (2.3)   mence work on that date. In effect, it appears that the contractor would be entitled to take a leisurely approach to starting work provided that it finished by the contract completion date. Of course, in practice, the contractor could be in

6.4.1.2   danger of having its employment terminated for failure to proceed regularly

110   and diligently with the Works.

12.3   Failure to give possession

If the employer fails to give the contractor possession so that it can begin the Works on the stated date, it will be a serious breach of contract. The contractor

111   will have a claim for damages and the completion date will cease to be operative. In such circumstances the contractor's obligation would be to complete within a reasonable time. Although that does not mean that the contractor has unlimited time in which to carry out the Works, it does mean that there is no date from which liquidated damages can run and, therefore, they are not deductible. If the employer's failure to give possession lasts more than a few days, it seems that the contractor may well have grounds to consider the employer's breach as an intention to repudiate the contract. The contractor should serve notice on the employer (see Box 12.1). The contractor's reference to 'time at large' will be noted. That phrase is given to the situation when, for any reason, there ceases to be a fixed date for completion of the Works. What the contractor is saying is that when possession is not given on the due date, the date for completion cannot apply and the contractor has a reasonable time in which to complete the Works; (see also Chapter 13). In any case, failure to give possession clearly requires negotiation between the employer and the contractor to achieve an amicable settlement and the architect should remind the employer of the duty to give possession before the date for commencement (see Box 12.2) and give advice if the employer is in breach (see Box 12.3).

Box 12.1   Letter from contractor to employer if possession not given on the due date.

SPECIAL DELIVERY POST

Dear Sir

PROJECT TITLE

May we draw your attention to the fact that possession of the site should have been given to us on [*insert date*] to enable us to commence the Works in accordance with clause 2.2 [*substitute '2.3' when using MWD*] of the conditions of contract. Possession was not given to us on the due date.

This is a serious breach of contract for which we will require appropriate compensation. It also places time at large and our obligation is now to complete within a reasonable time. We reserve all our rights and remedies in this matter. Without prejudice to the foregoing, we suggest that a meeting would be useful and we look forward to hearing from you.

Yours faithfully

Box 12.2   Letter from architect to employer before date for possession.

Dear Sir

PROJECT TITLE

Would you please note that the contractor is entitled to take possession of the site on the [*insert date*].

Will you please make sure that everything is ready so that the contractor can take possession on the due date? Failure to give possession on the due date is a serious breach of contract which can have serious consequences: the contractor may be able to claim substantial damages or even, if the delay is prolonged, treat the contract as repudiated. Moreover, the contractor's obligation to complete by the contract completion date may be removed.

Please let me know immediately if you anticipate any difficulties.

Yours faithfully

Although the architect's power to instruct the contractor probably extends to postponing the work, postponement is not the same as failure to give possession. If the architect postpones the work, the contractor will have possession of the site, but the carrying out of the Works will be suspended. The contractor may well wish, and it is entitled, to use the time to reorganise its site arrangements, repair site offices, improve its security, etc.

The contractor's licence to remain on the site ends on the date of practical completion but it has a restricted licence to continue to enter the site to deal with such

2.10 (2.11)   defects as are notified to it as defects in the rectification period (see section 20).

> **Box 12.3   Letter from architect to employer if there is a failure to give possession on the due date.**
>
> Dear Sir
>
> PROJECT TITLE
>
> I understand/have been notified by the contractor [*delete as appropriate*] that you were unable to give possession of the site on the date stated in the contract as the 'Works commencement date'.
>
> You will recall that, in my letter dated [*insert date*] I advised that failure to give possession on the due date is a serious matter.
>
> I am afraid that you must negotiate with the contractor if you wish to avoid the allegation of repudiation and possibly substantial damages. If you wish, you can instruct me to negotiate on your behalf but because this is outside the terms of my appointment, I will require your written authorisation to act for you in this way and your agreement to pay my additional fees and costs on a time basis as set out in my original conditions of appointment. I should also make clear that I can give no guarantee of the outcome of the negotiations.
>
> Yours faithfully

12.4   Common problems

Can the contractor insist that the architect physically sets out the building on site?

Unlike other JCT contracts, there is nothing in MW and MWD which says that the contractor must set out the Works on site. Therefore, it is always a good idea for the specification to make clear that it is the contractor's responsibility. However, if there is nothing in the specification, the contractor will be responsible for setting out the building accurately if the architect has supplied drawings which enable the contractor to do so. That is because, under the terms of the contract, the contractor has undertaken to carry out and complete the Works in

2.1.1 (2.1)   accordance with the contract documents.

The contract could have, but does not, say that the contractor gets possession of the site. It merely says that the contractor may commence work on a specific date. Therefore, can the employer carry on occupying the house or whatever relatively small building is concerned?

It is unfortunate that the contract does not expressly give the contractor possession of the site. Nevertheless, possession will be implied under the general law. If the project is a new building, there will only be the site to occupy and that will not usually be a problem. Most problems occur when the project involves work to an existing house and the owner either refuses to leave while work proceeds

112

or, having left, returns before practical completion and the contractor has to work in difficult circumstances. If there is nothing in the contract or in the specification to indicate the degree of possession to which the contractor is entitled, it will be implied that, at the very least, the contractor must have sufficient possession to carry out the Works without the employer getting in the way. The general principle is this: if it can be shown that the contractor reasonably expected to have a certain degree of possession, that is the degree of possession which must be given. If the employer at any time takes back partial possession or refuses to give up possession, the question will be whether the contractor is prevented or hindered in carrying out the Works by the unexpected presence of the Employer. If the answer is 'yes', the employer will be in breach of contract and the contractor will be entitled to claim damages through the courts.

13   Extension of time

13.1   Why necessary?

Every architect and contractor knows that the contract allows the architect to extend the time for completion of the Works in certain circumstances. Not everyone knows why. Therefore, it may be helpful to explain that first because a clear understanding will assist both architect and contractor to operate this process correctly. If everyone knows what they are doing and most importantly why, some of the common misconceptions and disputes may hopefully be avoided.

An example

Let us assume that there is no provision for extending time, but all the rest of the contract is as it is. There is still a date for commencing the Works and a date for completing them. Take the case of a contract to build a house. Now assume that, after the contractor has been building happily for a few months the employer wants some significant changes to the interior, for example having a large room instead of two smaller ones and having suspended timber ground floors instead of solid concrete with underfloor heating and so on. The architect issued an instruction and the contractor takes the instruction into account in the construction. It will soon be clear that the contractor will not be able to finish by the completion date, but a contract has been signed by which the contractor agreed that it would finish by that date. Is the employer entitled to deduct liquidated damages if the contractor is late due to the instruction? That does not seem fair or reasonable or sensible.

The answer

The law says that if the employer does something to prevent the contractor from finishing by the agreed completion date, the contractor is no longer obliged to finish by that date and it is entitled to finish within a reasonable time. Who decides what a reasonable time might be? A judge or arbitrator or adjudicator if there is a dispute about it. This situation is what is meant when a contractor says that 'time is at large'. Note that time would become at large even though the architect was doing something which the contract permitted to be done, in this case: issuing an instruction to vary the Works. Obviously, there are many things for which the employer or architect is responsible which could stop the contractor finishing on time.

Extension of time clauses are included in building contracts to prevent time becoming at large. They give the architect power to extend time for delays which are the result of actions or inactions on the part of the employer or the architect as well as for other reasons for which neither employer nor contractor are responsible. The proper exercise of this power by the architect should prevent time becoming at large. This is important because liquidated damages (which will be explained in the next chapter) cannot be recovered unless there is a definite date for completion. In that case, the employer would be obliged to try to recover ordinary unliquidated damages by suing the contractor for breach of contract and then trying to prove the amount lost. An architect who fails to give an extension of time which is properly due to the contractor may be held to be negligent.

13.2 Extension of time

In the absence of an extension of time clause, neither the employer nor the architect would have any power to extend the contract period. MW and MWD both entitle the architect to extend time. It cannot be over-emphasised that the only effect of the architect giving an extension of time is to fix a new date for completion and, incidentally, to relieve the contractor from paying liquidated damages at the stated rate for the period in question. It certainly does not automatically entitle the contractor to payment of additional preliminaries, as many contractors, architects and quantity surveyors believe. More about that later.

The date for completion is either the date stated in the contract particulars or the date which the architect has later fixed in response to a notification by the contractor. If during the progress of the Works it becomes clear that the contractor will not be able to complete them by the completion date, the contractor must immediately notify the architect in writing. If the delay is *for reasons beyond the control of the contractor*, including compliance with any instruction of the architect the issue of which is not due to a default of the contractor, the architect must then make, in writing, a *reasonable* extension of time. The contract does not say when the architect must do this, but the extension of time must be made as soon as possible. Certainly the application should not be put on one side and must be made before the current date for completion is passed, if practicable.

It used to be thought that it was sufficient for the architect to form an impression of the period; that is not correct: the period must be calculated. It is essential that the architect establishes that the Works will not be completed by the date for completion in the contract or any later extended date before an extension of time is given. Because the contractor has suffered a delay amounting to three weeks does not automatically entitle it to three weeks' extension of time. It is the effect of the three weeks on the completion date that is the crucial factor.

Figure 13.1 illustrates the contractor's duties in claiming an extension of time. Figure 13.2 sets out the architect's duties in relation to such a claim.

The RIBA publish a form which may be used to give an extension of time. It is sensible to use this form, although a letter would serve just as well. The important thing is that the architect should not be tempted to go into great detail about the reasons for giving the extension of time.

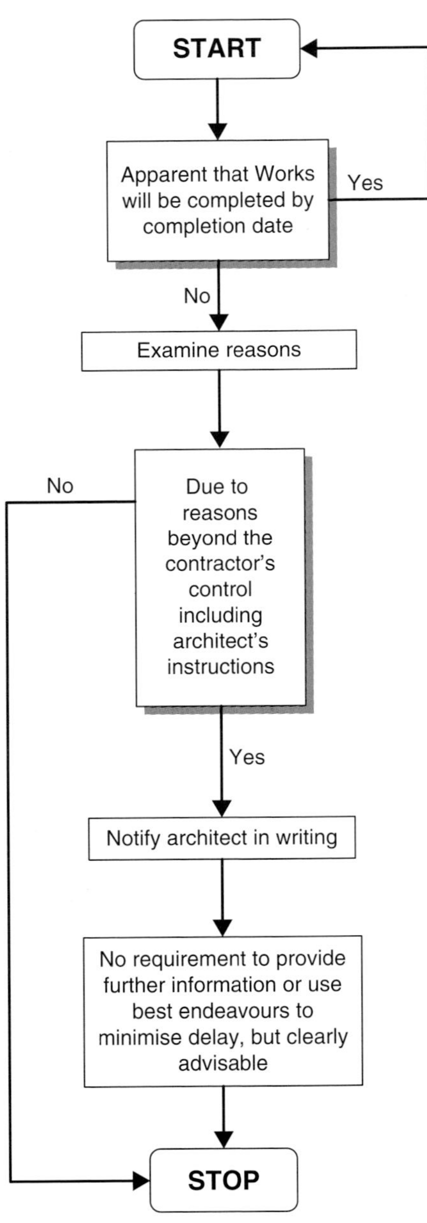

Figure 13.1 Flowchart of contractor's duties in claiming an extension of time.

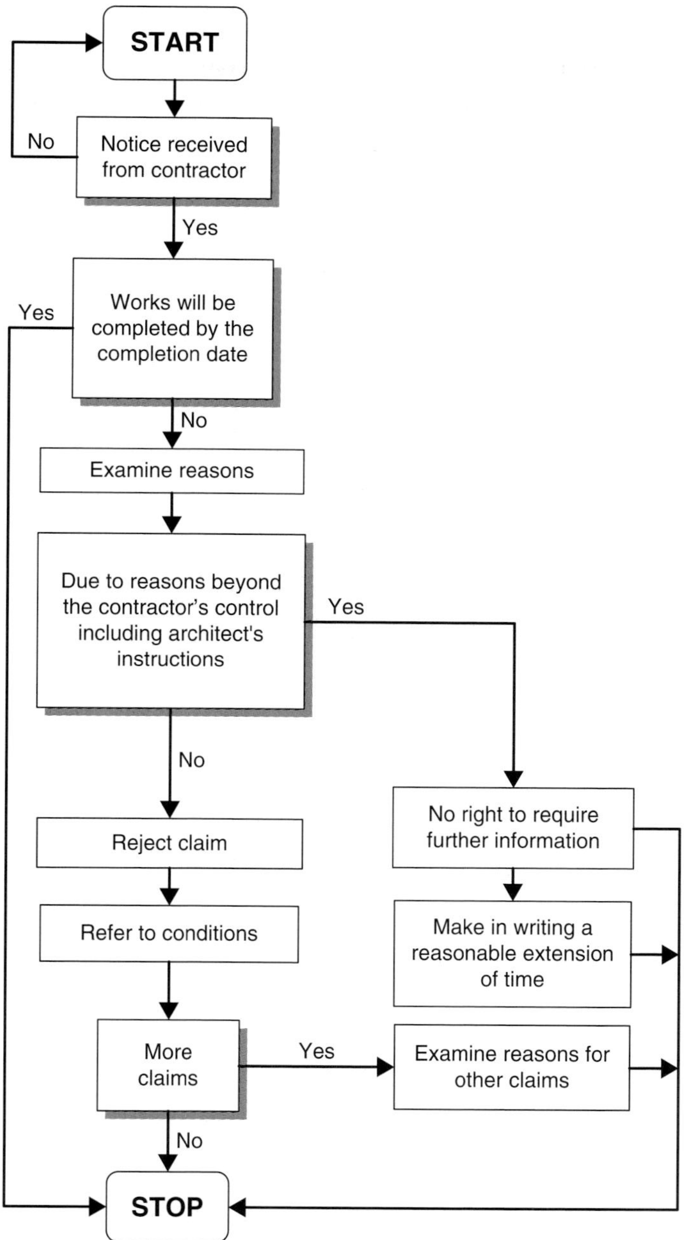

Figure 13.2 Flowchart of architect's duties in relation to claim for extension of time.

13.3 Reasons

The architect must of course be satisfied that the completion date will not be met because of 'reasons beyond the control of the contractor'. This phrase is very broad and it expressly includes compliance with architect's instructions but excludes instructions which are given as a result of some default on the

1.1

119

120

part of the contractor. The contract makes clear that defaults of the contractor and any of the 'Contractor's Persons' are within the contractor's control. 'Contractor's Persons' is defined as including all the contractor's employees, agents and anyone engaged on site in connection with the Works except of course the architect, statutory undertakers, the employer and anyone authorised by the employer. It is now clear that sub-contractors and suppliers are within the control of the contractor. It is not clear whether 'reasons beyond the control of the contractor' in fact extends to all delays which are the fault of the employer. Sensibly, it ought to do so, but the case law suggests otherwise. The contractor is entitled to an extension of time if the delay is caused by inclement weather. It need not be 'exceptionally adverse weather but only the kind of weather which delays the contractor. There is no test to be applied such as whether the contractor ought to have allowed for it given the time of year.

The contract does not actually require any reasons to be stated when the architect gives an extension of time. Giving detailed reasons is simply inviting the contractor to take issue with the architect's calculation. Failure properly to give an extension of time may result in the contract completion date becoming at large, and liquidated damages becoming unenforceable.

13.4   Failure to notify delay

121

What happens if the contractor fails to notify the architect that the completion date will not be met? Clearly the contractor is in breach of contract in not doing so because the obligation rests on the contractor, and it is suggested that the contractor's failure to give notice is a matter which may be taken into account by the architect in determining the extension of time. In taking account of the failure, the question to be asked is whether the contractor's failure prejudiced the employer in any way. The critical date is the date it became apparent that the Works would not be completed on time. In other words, if the architect had been informed immediately, could any measures have been taken to reduce or eliminate the delay? It is certain that the contractor's notice is not a pre-condition to the giving of an extension of time and it is up to the architect to monitor progress and in an appropriate case award an extension of time even if the contractor has not given any notification. As explained at the beginning of this chapter, the primary purpose of an extension of time clause is to protect the employer against the loss of liquidated damages following some act of hindrance or prevention. The contractor's duty is to notify all delays whether or not they may in principle entitle the contractor to an extension of time. Therefore, the contractor must also give notice about its own delays so that the architect can monitor the situation and issue instructions if appropriate.

13.5 Does an extension of time entitle the contractor to any money?

That simple answer to that is 'No'. It is traditional in the construction industry for claims by the contractor for both extra time and extra money to be linked together. However, there is no necessary link between time and money. An extension of time is not a pre-condition to a monetary claim for loss and/or expense under any of the JCT Forms, neither does an extension of time automatically entitle the contractor to loss and/or expense. The extension of time provision is purely to give the contractor an extension to the period for carrying out the Works.

2.7 (2.8)

13.6 Common problems

The contractor was late in getting the roof covered and a two-day downpour prevented the contractor from dong any work at all.

Many architects, if confronted with this type of claim from the contractor for a two-day extension of time, would reject it on the basis that the contractor caused the delay due to its failure to comply with its own programme. That would be wrong. The test is whether the cause of the delay (the rain) was within the contractor's control: clearly not. Therefore, an extension should be given. The contractor had no strict duty to comply with its programme. If the Works as carried out are delayed by weather, an extension of time is indicated.

122

The contractor informs the architect that the employer has called on site and agreed an extension of time. On enquiry, the employer confirms it.

Only the architect may give an extension of time under the terms of the contract. The employer and the contractor are entitled to agree a new date for completion, because they are the parties and they can agree whatever they choose. However, in doing so, the employer is not giving an extension of time; the parties are varying the contract terms.

Although the end result may be the same, the means for achieving it are completely different. That is because what the employer and the contractor are doing is to enter into a new little contract to vary the old. That little contract must satisfy the requirements for a binding contract if it is to have any effect. By reference to Chapter 1 it can be seen that both parties must give something (consideration). So far as the employer is concerned the giving involves allowing the contractor to complete later and foregoing the liquidated damages that would otherwise be due for late completion. It is not at all clear what the contractor is giving. In these circumstances, there would be no little contract and the employer's agreement is void or at least voidable if either contractor (unlikely) or the employer suffers a change of mind. This kind of ad hoc arrangement between employer and contractor is unfortunately all too common.

The bottom line is that the employer should not make any kind of agreement with, or give any instructions to, the contractor. Every communication with the contractor should be through the architect or, at the very least, with the architect present. Otherwise the architect is put into a position where contract administration becomes impossible.

14   Liquidated damages

14.1 What are liquidated damages?

They are the loss which the employer may suffer as a result of the contract Works being delayed beyond the completion date. They are called 'liquidated' to distinguish them from 'unliquidated' damages. Unliquidated damages have to be proved before they can be recovered. If there was no liquidated damages provision in the contract and the contractor failed to complete by the completion date entirely due to its own fault, it would be in breach of contract. The employer would have a claim for whatever damages the employer suffered as a result. In order to get those damages, and assuming that the contractor did not just pay on demand, the employer would have to use one of the dispute resolution procedures in the contract. Then the employer would have to prove that the contractor had a duty to complete by the date for completion in the contract and that the contractor failed to do that through its own fault. Having established that, the employer would then have to set about proving exactly how much had been lost as a direct result of the contractor's breach of contract. This is all very time consuming and expensive. To avoid that, the contract includes a provision for liquidated damages. The idea is that they are a fixed sum per week or day or whatever period is agreed. If the contractor delays completion and there is no extension of time to cover the delay, the employer is not faced with proving the loss; the assumed loss is already agreed.

Principles

There are many misconceptions about liquidated damages. The most important point is that the sum stated as such is recoverable whether or not the employer can prove that any loss has been incurred as a result of late completion or even if it turns out that no loss at all has been suffered. On the other hand, the employer cannot recover more than the amount of liquidated damages even if the employer actually suffers a much greater loss. It used to be said that the object of the liquidated damages clause was to fix an amount which was a pre-estimate of the amount of damages which the employer might suffer if the contractor was late in completing. This view has been recently modified so that, in relation to construction contracts, the amount should be a sum which is not exorbitant when viewed in regard to the importance to the employer of getting the project completed on time. It is irrelevant to consider whether in fact there is any loss.

14.2   Liquidated damages or penalty?

Contractors often refer to the liquidated damages provision as 'the penalty clause'. Strictly speaking that is incorrect. In law a penalty is not enforceable. A sum is treated as liquidated damages (and so is recoverable) if it is a fixed and agreed sum which is not exorbitant when viewed in regard to the importance to the employer of getting the project completed on time, or a lesser sum, estimated at the time the contract is made. It does not matter if the estimate is a poor one. A penalty is not related to probable damages but may be much more than that. Deciding whether a sum is liquidated damages or a penalty can be difficult where the clause is complicated. However, the courts have indicated that, in deciding the issue, they will not take account of hypothetical situations. They are much more likely to take a pragmatic approach. It is even acceptable to express the liquidated damages as a series of graduated amounts related to the seriousness of the breach.

Calculation

Under MW and MWD the amount inserted as liquidated damages is usually relatively small to reflect the potential loss to the employer from late completion, but it is bad practice to pluck a figure out of the air. The employer, in consultation with the architect, should have calculated the amount of liquidated damages carefully at pre-tender stage.

To set the figure at the right level, the employer should discuss it carefully with the architect before making the calculation. This is sometimes difficult. In the case of profit-earning assets, there is no problem. All that then need be done is to analyse the likely losses and additional costs. The following should be considered:

- Loss of profit on a new building, e.g. rental income, retail profit, etc.;
- Additional inspection and administrative costs – including additional professional fees;
- Any other financial results of the late completion, e.g. storage charges for furniture in the case of a domestic building, renting alternative accommodation.

The figure thus arrived at must be inserted in the contract particulars. However, liquidated damages are all that the employer can claim for late completion; therefore, it is important that everything that could arise is included in the calculation If no figure was inserted, no liquidated damages would be payable. The inclusion of 'Enil' would amount to the same thing, because the rate would be

127

128

2.8 (2.9)

2.9 (2.10)

129

£nil per week. The same result would follow if no completion date was inserted because obviously there must be a date from which liquidated damages can be calculated.

Liquidated damages are payable at the specified rate only if the Works are not completed by the original completion date or extended contract completion date. They are payable by the contractor at the stated rate per week for the period between the stated completion date and the date of practical completion as certified by the architect. The contract allows the employer to deduct liquidated damages from monies due to the contractor, e.g. interim payments, or the employer may recover them as a debt.

The employer must be careful not to lead the contractor to believe that liquidated damages will not be deducted. This is sometimes done when no extension of time is due, but the employer does not want to frighten the contractor with the prospect of heavy damages. In these circumstances the employer may assure the contractor that liquidated damages will not be deducted. However, an employer who does make such representations may be estopped (prevented) from later having a change of mind and deciding to recover the liquidated damages after all. This will be particularly the case if the contractor, relying on the employer's assurances, has paid out money to sub-contractors without any deductions.

14.3   Procedure

2.8 (2.9)

2.8.1 (2.9.1)

2.8.2 (2.9.2),
4.5.4

It is important that the employer gets the process right and does everything that needs to be done in order to recover the liquidated damages. Although the architect is not in a position to advise the employer whether or not to recover the damages, the architect should draw the employer's attention to the right to recover damages and what the contract requires to be done.

If the contractor fails to complete the Works by the completion date, the employer is entitled to start recovering liquidated damages and may continue until practical completion. The employer can of course decide not to recover the damages. The contract does not expressly permit the employer to recover something less than the rate in the contract particulars, but the contractor is unlikely to complain. However, the employer cannot recover at a rate greater than the rate in the contract particulars. If the employer chooses to recover the damages, the choice is between deducting the damages from payments due to the contractor or to ask the contractor to pay the damages. In practice, the employer will usually deduct, because it is easier for the employer to hold back money than to get the contractor to pay it.

If the employer intends to deduct, a pay less notice must be issued no later than five days before the final date for payment giving details of the calculation of the deduction based on the rate and the period of delay (see Chapter 18). In addition, if the employer intends to deduct the damages from the amount in the final certificate or even if the recovery is to be made as debt from the contractor, an extra notice must be issued to the contractor before the date the final

<table>
<tr><td></td><td>certificate is issued. It should be noted that this is in addition to and does not</td></tr>
<tr><td>2.8.3 (2.9.3)</td><td>replace the pay less notice.</td></tr>
</table>

14.4 Common problems

If the contractor's employment is terminated after the date for completion, can liquidated damages be deducted for every week until the project is completed by another contractor?

Once the contractor has left the site, it can no longer complete the Works. Therefore, there can be no practical completion under the contract. If the employer engages another contractor to finish the Works, that will be under a different contract and when practical completion is eventually certified, it will not relate to the original contract. Therefore, once the contractor's employment has been terminated either liquidated damages run on *ad infinitum* or they stop at that point. The right of the employer to terminate is incompatible with the employer's right to deduct liquidated damages after termination.

130

The contractor says that the employer cannot charge for having to correct defective work after practical completion, because the amount of liquidated damages is the maximum that the employer can deduct.

The contractor is confused in this instance. Liquidated damages is the maximum amount that the employer can deduct to compensate for late completion. However, defective work, whenever it may be discovered, is nothing to do with late completion and, if the contractor fails to rectify it, the employer is entitled to get another contractor to carry out rectification work and recover the cost from the original contractor.

15 Financial claims

15.1 General

It is well understood that the contractor is entitled to be paid the amounts included by the architect in interim certificates up to the value of the Contract Sum. The contractor is also entitled to have added to the Contract Sum the value of any variations to the Works. Sometimes, of course, a variation may result in a reduction to the Contract Sum. In addition to those sums, it is recognised that the contractor may incur other expense or losses for various reasons. For example, the architect may instruct a variation to the Works and simply valuing the variation in accordance with the contractor's rates in the priced document may not adequately reimburse it, possibly because the variation has to be done out of sequence or it delays other work. Other costs to the contractor may arise because the architect has been slow to provide drawings or give instructions or because the employer has failed to allow the contractor to start on time or has stayed in a property after agreeing to move out. The actions or inactions of the employer or architect have the potential to cause the contractor expense or loss.

MW and MWD are unique among the JCT contracts in that they contain no clause entitling the contractor to direct loss and/or expense except for the provision in regard to the valuation of variations which includes any direct loss and/or expense incurred by the contractor due to regular progress of the Works being affected by compliance with a variation instruction. The contractor cannot 'claim' loss and/or expense on its own initiative.

3.6.3

It is traditional in the construction industry for claims by the contractor for both extra time and extra money to be linked together. However, it should be borne in mind that there is no necessary link between time and money. The grant of an extension of time is not a pre-condition to a monetary claim for loss and/or expense under any of the JCT forms. An extension of time does not automatically entitle the contractor to loss and/or expense.

Direct

The contract refers to 'direct' loss and/or expense to make clear that the contractor is not entitled to indirect loss and/or expense. Therefore, the kind of

The JCT Minor Works Building Contracts 2016, Fifth Edition. David Chappell.
© 2018 John Wiley & Sons Ltd. Published 2018 by John Wiley & Sons Ltd.

loss and/or expense which the architect is entitled to include in the valuation of a variation is what anyone would have understood would be the result of the variation and the variation must have directly caused it without anything else intervening. For example, a late variation to brickwork when most of it has been constructed might entail the demolition of most of it and a long delay while the contractor carried out the variation. However, take the situation where the contractor has scaffolding in place and has almost finished the roofing when the architect instructs the insertion of a dormer window. During the construction of the dormer window, the scaffolding collapses causes a great deal of damage to work already constructed. The contractor may argue that the collapse was due to the architect's instruction, because had it not been issued the scaffolding would have been dismantled and removed by the date of the collapse. But that instruction could not have been the direct cause of the collapse. The direct cause might be that it was not properly erected in the first place and during use a vital component worked loose or it may have been vandals. The instruction resulted in the scaffold being in place longer but did not result in its collapse. The contractor would probably be entitled to some extension of time due to the delay caused by the construction of the dormer, but not the extra time involved in rebuilding the scaffolding and repairing the damage.

15.2   Dealing with loss and/or expense

When valuing a variation, the architect must include any direct loss and/or expense caused by the instruction. Probably, the first thing for the architect to do is to ask the contractor for evidence that some loss and/or expense has been suffered due to regular progress being affected. The usual situation is that after the issue of the instruction, the contractor will notify a delay and ask for an extension of time. If an extension of time is given for the delay caused by the instruction, it makes sense for the architect to use that period as the measure of the loss and/or expense. Even if no delay to the completion date results from the instruction, that may be simply because the activity affected was not on the critical path. In such a case, there would be no extension of time, but regular progress may be affected in the non-critical activity and the architect must try to determine what loss and/or expense the contractor has suffered.

The contractor will have its own ideas about how the loss and/or expense is to be ascertained. Under the SBC contracts, the architect may instruct the quantity surveyor to ascertain the amount of money due, but under MW and MWD there is no quantity surveyor and the task is left to the architect. So how does the architect set about the task? Let us assume that the architect has asked the contractor for evidence of the loss and/or expense and the contractor has provided its idea of what it should be paid. The architect must go through the information and see if it is convincing. It is worth remembering that the contractor is not entitled to some kind of windfall in this situation but only what can be shown to be the amount the contractor has had to spend, in excess of what it is being valued for the work and materials, or has lost due to the instruction.

The contractor must provide evidence of the actual amounts and not estimates or figures derived from formulae.

If the architect has given an extension of time which can be used as a basis, Table 15.1 lists the kinds of things which the contractor might say constitute the loss and/or expense. The list is not exhaustive and it is a matter for the architect to decide whether the items put forward by the contractor are valid. The comments are pointers rather than detailed expositions of the legal position.

Table 15.1   Common loss and/or expense items.

Item	Comments
Onsite overheads such as site accommodation, health and safety facilities, tools, telephone, electricity, welfare and sanitary facilities.	These constitute what would normally be termed 'preliminaries' and there may be a temptation to simply use the amount per week which the contractor has allowed. However, this assumes that the contractor correctly calculated what the preliminary costs would be. These costs could be used if both contractor and employer agree, but loss and/or expense should be the actual loss and actual expense proved by documents.
Site supervision being extended.	It is relatively unusual for a small project to warrant a dedicated site agent. Usually there is a working foreman and the contractor has a travelling supervisor for all the projects in the area. If the contractor claims for a dedicated foreman during the delay period, it must be shown that the foreman is the regular foreman and not simply included in the costs of the delay because the contractor had a spare foreman at the time.
Plant, vehicles and equipment	If the contractor has hired in plant etc, it is entitled to be paid the hire charges, but only during the period of the delay which is not the amount at the end of the project but the amount of delay when the delay occurred. If the plant etc. is not hired in, all the contractor can usually expect is the depreciation value.
Head office overheads	In order to be able to claim this, the contractor has to show that it actually turned down other work because its workforce was tied up on this project after the date for completion. This is a claim for loss of opportunity to earn a contribution to the head office overheads.
Increased costs caused by the delay.	The contractor would have to prove the increases and not rely on formulae. It would also have to show that it did what it reasonably could to avoid those costs, such as ordering on time.
Cost of extending the works and other insurances.	This is a valid claim subject to proof.

How the contractor can claim

Although there is limited scope for the architect to include direct loss and/or expense in a valuation instruction, there is no clause in the contract which entitles the contractor to make loss and/or expense claims against the employer whether arising from prolongation or disruption. This leads some people to suppose that this is a risk which the contractor must price. Nothing could be further from the truth. The absence of a direct loss and/or expense clause merely means:

- There is no contractual provision for ascertaining and paying money claims.
- The architect has no power to quantify or agree such claims.

The absence of a claims clause is not necessarily a benefit as some employers appear to think, because it deprives the architect of the power to deal with all kinds of money claims through the contract mechanism. That power can be very useful. But the absence of the power should not mislead either employer or architect into thinking that the contractor cannot make loss and/or expense claims for anything other than variation instructions. It can; but these other claims can only be made outside the contract terms by way of adjudication, arbitration or litigation as appropriate using the common law. They must be based on the employer's breach of some express or implied term of the contract or some other legal wrong. A favourite implied term relates to co-operation and non-interference and that the employer:

- Would do all things necessary on their part to enable the contractor to carry out and complete the Works expeditiously, economically and in accordance with the contract.
- Conversely, that neither the employer nor the architect would in any way hinder or prevent the contractor from carrying out and completing the Works expeditiously, economically and in accordance with the contract.

It is suggested that these terms are to be implied when the contract is in MW or MWD form and this being so opens up a wide area for common law claims.

The architect has no power to deal with these claims, unless the employer expressly authorises the architect to do so, and the contractor agrees. Even then, the architect cannot certify payment of these kinds of claims, which must be settled outside the terms of the contract.

15.3   Types of claims

There are four type of claims which an architect ought to know about – at least the basics. They are as follows:

- Contractual claims: claims made under the terms of the contract and with which the architect may deal. MW and MWD do not include for this kind of claim but other contracts usually do so. This is the only kind of claim which the architect is entitled to consider.

- Common law claims, sometimes called ex-contractual: claims made outside the terms of a contract relying on the general law.
- *Quantum meruit* claims: claims which can be made if no price has been agreed.
- *Ex gratia* claims: claims which have no legal basis but which a contractor may make if losing money. The employer has no legal obligation to pay this kind of claim.

Table 15.2 summarises the MW and MWD clauses which may give rise to claims.

Table 15.2   Clauses that may give rise to claims under MW and MWD.

Clause	Event
2.2 (2.3 under MWD)	Failure to allow commencement on the due date by lack of possession or otherwise
2.3 (2.4 under MWD)	Failure to issue further necessary information
2.4 (2.5 under MWD)	Errors or inconsistencies in the contract documents
2.5 (2.6 under MWD)	Divergence between statutory requirements and contract documents or architect's instructions
2.7 (2.8 under MWD)	Failure to give extension adequately or in good time
2.8 (2.9 under MWD)	Wrongful deduction of damages
2.9 (2.10 under MWD)	Failure to certify practical completion at the proper time
2.10 (2.11 under MWD)	Wrongful inclusion of work not being defects etc.
2.11 (2.12 under MWD)	Failure to issue certificate that the contractor has discharged its obligations
3.1	Assignment by employer without consent
3.3.1	Unreasonably withholding consent to sub-letting
3.4	Failure to confirm oral instructions
	Instructions altering the whole character or scope of the work
	Wrongful employment of others to do work
3.6	Variations
	Wrongful omission of work to be done by others
3.8	Unreasonably or vexatiously instructing removal of employees from Works
4.3	Failure to certify payment in accordance with the contract
	Failure to certify payment for materials properly on site
	Failure to certify payment at practical completion
4.8.2	Failure to issue final certificate
4.10	Fluctuation provisions
6.4	Invalid termination
	Payment after termination
6.5	Invalid termination
6.7	Payment after termination
6.11	Payment after termination

Note: There is no loss and/or expense clause in this contract. All claims have to be made at common law.

The following book will be found useful for architects, contractors and sub-contractors wishing to be better informed about the complex matter of claims: *Building Contract Claims*, 5th edition, by David Chappell, Wiley-Blackwell, 2011.

15.4   Common problems

The contractor says that it is entitled to preliminaries on the two weeks extension of time it has been given for late provision of architect's details

It is a common misconception among architects, contractors and some quantity surveyors that an extension of time entitles the contractor to extra preliminaries. The arguments goes like this: the contractor had a weekly sum in his price for preliminaries at, say, £1,000 per week. The architect, by extending the time for completion has admitted that the contractor needs the additional two weeks because the architect was late in the provision of details. The contractor's price breakdown shows that it is costing it £1 000 for every week it stays on site and, therefore, the preliminaries must be automatically extended by £2,000. That is fair and reasonable.

3.6.3

Well it may or may not be fair and reasonable, but that is not the point here. The point is that there is no provision in the contract for the contractor to recover its preliminary costs as a result of getting an extension of time. The only thing that an extension of time gives the contractor is more time. The cause of the delay in this instance was the architect's late provision of details. Therefore, the architect cannot give any loss and/or expense (which is what the extra preliminaries actually is) under the contract which only allows loss and/or expense to be added to the valuation of an architect's variation instruction. If the contractor is seeking the £2,000, it can only recover it by making a claim for breach of contract against the employer using the contractual dispute resolution procedures. The problem for the contractor is that the cost and risks of going through a formal procedure to recover far outweigh the £2,000 being claimed.

A final thought: the £1,000 per week in the contractor's price is not the cost to the contractor of each week on site, it is what the contractor is charging the employer each week. The contractor's cost must be less than that or the contractor will soon go out of business.

The architect has agreed a price with the contractor for an instructed variation. Now the contractor is saying that the contract requires the architect to add some loss and/or expense to the agreed price.

3.6.2
3.6.3

The contract requires the architect and the contractor to endeavour to agree a price for each variation. If there is a failure to agree, the architect must carry out the valuation and include any direct loss and/or expense incurred by the contractor.

Not only would it make no sense if the price agreed between architect and contractor did not include loss and/or expense incurred, it is clear from the contract that it must be included in what is agreed. There are two stipulated methods of arriving at the price: agreement between contractor and architect or, failing that, valuation by the architect. It is clearly intended that both procedures should arrive at a comparable figure.

16 Architect's instructions

16.1 Architect's instructions

All instructions to the contractor must be confirmed in writing in order to be effective. Despite what is often thought, there is absolutely no requirement that instructions must be on a specially printed form headed 'Architect's Instruction', although it is undoubtedly good practice to use such forms because:

- It leaves no room for doubt that the architect is issuing an instruction. An instruction buried in a letter which deals with a great many other things can sometimes be overlooked.
- It makes the job of keeping track of instructions for the purposes of valuation and checking so much easier.

Instructions can be given in letter form, provided it is made clear that the architect is issuing an instruction (Box 16.1). Hand-written instructions can also be given on site, provided they are signed and dated (the architect should always sign on behalf or in the name of the architect named in the Articles). Instructions contained in the minutes of site meetings are valid if the architect is the author of the minutes and if they are recorded as agreed at a subsequent meeting. It is probable, however, that such instructions are not effective until the contractor receives a copy of the minutes recording agreement. Since site meetings are sometimes at monthly intervals, the matter giving rise to the instruction may be ancient history before the contractor has a duty to comply; minuted instructions are, therefore, best avoided.

The position with regard to drawings is uncertain. If the architect issues a drawing together with a letter instructing the contractor to use the drawing for the Works, that is certainly an instruction for contract purposes. If the drawing is simply issued under cover of a compliments slip, the drawing may be an instruction or it may simply be sent for comment; in most cases it will be taken to be an instruction. The contractor would be wise to check first, but the architect may have difficulty showing that the drawing was not intended as an instruction if the contractor simply carries out the work. Note that if the architect hands the contractor, under cover of a compliments slip, a copy of the employer's letter requiring some action, it is probably not an instruction at all although

> **Box 16.1   Letter from architect to contractor issuing an instruction.**
>
> Dear Sir
>
> PROJECT TITLE
>
> In accordance with clause [*insert the relevant clause*] please carry out the following instruction[s] forthwith:
>
> [*insert the instruction*]
>
> Yours faithfully

most contractors would take it as an instruction. The moral is clear: compliments slips should not be used where an instruction is intended.

3.4

The issue of instructions is covered in general by a very broadly worded provision in the contract. The procedure is shown in Figure 16.1. Despite what, at first sight, may appear to be an all-embracing provision ('The Architect… may issue written instructions and the Contractor shall forthwith comply with them…') the architect must act within the scope of authority and the instruction must relate to the Works. The courts do not generally favour excessively broad clauses and prefer to see respective rights and duties expressed in precise terms. In the event of a dispute, therefore, it is likely that the clause will be given a very narrow interpretation restricting it to instructions mentioned elsewhere in the contract or instructions which are obviously essential. The contractor must carry out the instruction 'forthwith', i.e. as soon as it

132   reasonably can.

Oral instructions

Although people commonly refer to 'verbal' instructions when they mean instructions given by word of mouth rather than in writing, 'verbal' does not mean that. It strictly means 'of words'. The correct term is 'orally'. Provision is made for the situation if instructions are issued orally. They have no effect (in other words they are not architect's instructions) until they are confirmed in writing by the architect. The contract does not say when a confirmed oral instruction becomes effective. It is considered that, in context, the instruction becomes effective when the contractor receives the confirmation. Although the practice of issuing oral instructions is widespread, it is difficult to see why it was necessary to make special provision in the contract since the effect of the confirmation is no different from the effect of a simple written instruction in the first place. The oral instruction itself is irrelevant. Note that there is no provision for the contractor to confirm oral instructions. But if the contractor confirms an oral instruction from the architect and the contractor proceeds with the work, it

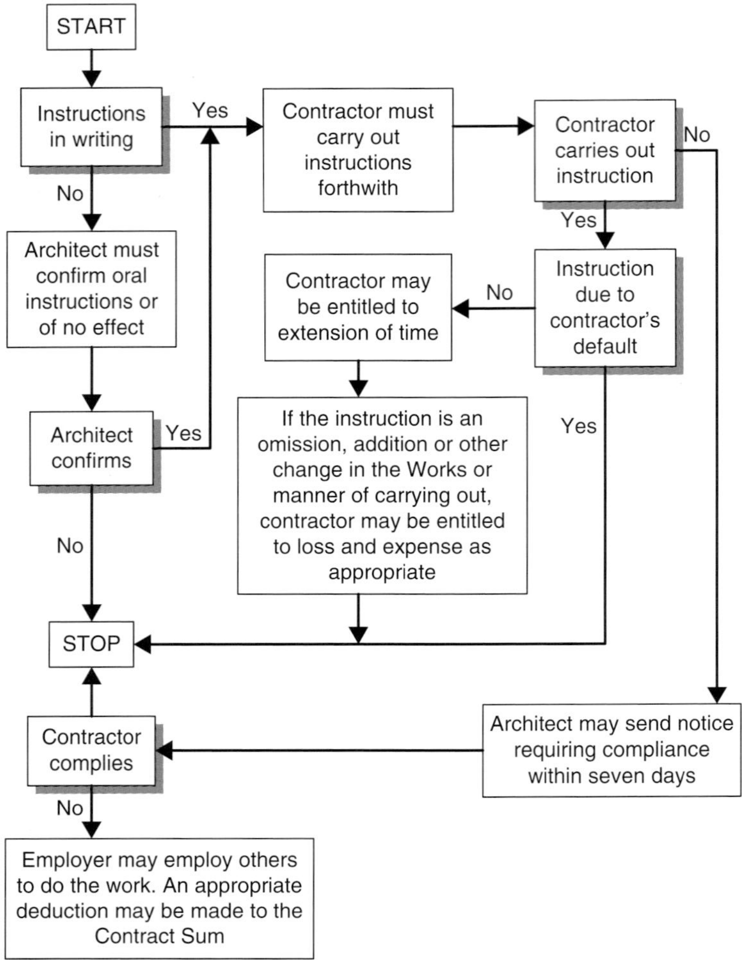

Figure 16.1 Flowchart of architect's instructions.

is possible that the employer will be legally prevented from denying the contractor's entitlement to payment. Ideally, oral instructions should be avoided.

If the contractor does not carry out instructions 'forthwith', the contract provides a remedy. As a first step, the architect must send the contractor a written notice requiring it to comply with the instruction within seven days (Box 16.2). If the contractor does not comply, the architect should advise the employer (Box 16.3) that others may be employed to carry out the work detailed in the instruction and the contractor will be liable for all the additional costs. Although the contract specifically states that it is the employer who may employ others, the employer will expect the architect to give advice and handle the details. It will amount to a completely separate contract and, in order to ensure that there can be no reasonable question about the costs to which the employer is entitled, the architect should obtain competitive quotations from three firms if time and circumstances permit. After the work is completed, the architect may make an

Box 16.2 Letter from architect to contractor requiring compliance with instructions before default action taken.

SPECIAL DELIVERY*

Dear Sir

PROJECT TITLE

On [*insert date*] I instructed you to [*specify*] under clause 3.4 of the Contract. That clause requires you to carry out my instructions forthwith.

You have failed to do so and in accordance with clause 3.5, I hereby require you to comply with the instruction to which I refer above. If you fail to comply within seven days after receipt of this notice, the Employer will engage others to carry out the work and an appropriate deduction will be made from the Contract Sum.

Yours faithfully

Copy: Employer

                   Quantity surveyor (if appointed)

Service by Special Delivery is not required by clause 3.5 but it is a desirable precaution.

Box 16.3 Letter from architect to employer if contractor fails to comply with instruction within seven days after notice.

Dear Sir

PROJECT TITLE

I refer to my letter to [*insert name of contractor*] dated [*insert date*] requiring it to comply with my instruction of the [*insert date*] within seven days.

The seven days expired yesterday and, during a site visit this morning, I observed that the Contractor had not complied with my instruction.

I advise that you may take advantage of the remedy afforded by clause 3.5 of the Contract and, if you will let me have your written instructions to that effect, I will obtain competitive tenders from other firms for the carrying out of the work. All additional costs, including my additional fees, will be deducted from the Contract Sum.

Yours faithfully

appropriate deduction from the Contract Sum in this regard in the next payment certificate.

Note that the 'additional costs' refer to additional costs over and above the cost of the instruction. It is not intended that the employer should get the instruction carried out at the contractor's expense.

The additional costs may include such items as scaffolding, cutting out and making good, depending on circumstances. The architect is usually entitled to charge additional fees and the architect may include them in the computation of costs together with any other incidental expenses attributable to the contractor's non-compliance. When preparing the final account, the contractor is entitled to a brief statement showing how the figure has been calculated.

Box 16.4 is the kind of letter the contractor may send when receiving a compliance notice and the contractor can always seek immediate adjudication on the matter.

16.2 Contractor's objection

3.4.2

3.6.1

Under MWD the architect is not entitled to issue an instruction which affects the design of the CDP Works without the contractor's consent. This prohibition is consistent with the architect's power to issue instructions requiring a change in the Employer's Requirements which result in an alteration to the design of the CDP Works. The point is that the architect may alter what the employer requires but it is a matter for the contractor to interpret that in terms of any alterations to the CDP design. However, the full extent of the prohibition is not entirely clear:

- It may mean that the architect can issue an instruction changing the Employer's Requirements without the contractor's consent even if that means the contractor having to change the CDP design but that the architect cannot issue an instruction directly altering the CDP design.
- Alternatively, it may be quite broad in scope and prohibit the architect from issuing either an instruction changing the Employer's Requirements which result in an alteration to the CDP design and also prohibit any instruction directly altering the CDP design, without the contractor's consent.

It seems clear that the architect has no power to issue an instruction directly altering the CDP design whether or not the contractor consents, but whether consent is required for an instruction changing the Employer's Requirements is a moot point. On balance it appears that the architect has no power to issue an instruction changing the Employer's Requirements without the contractor's consent. In practice, it is doubtful that any contractor will withhold consent provided that the change results in adequate additional payment.

Note that there is no general provision for the contractor to object to any instruction, or request the architect to state the empowering provision.

> **Box 16.4   Letter from contractor to architect on receipt of seven-day notice requiring compliance with instruction.**
>
> Dear Sir
>
> PROJECT TITLE
>
> We have today received your notice dated [*insert date*] which you purport to issue under clause 3.5 of the above Contract.
>
> [*Add either:*]
>
> We will comply with your instruction Number [*insert number*] dated [*insert date*] forthwith, but our compliance is without prejudice to and reserving any rights and remedies we may possess.
>
> [*Or:*]
>
> We consider that we have already complied with your instruction Number [*insert number*] dated [*insert date*]. If the Employer attempts to employ other persons and/or if you deduct any monies from the Contract Sum, this will constitute a serious breach of contract for which we will seek appropriate remedies. However, without prejudice to the foregoing, provided that you immediately withdraw your purported notice of non-compliance, our [*insert name*] will be happy to meet you on site or at our office to sort out what appears to be a simple misunderstanding.
>
> [*Or:*]
>
> It is not reasonably practicable for us to comply within the period specified by you because [*give reasons*]. You may be assured that we are well aware of our contractual obligations and intend to carry out your instruction Number [*insert number*] dated [*insert date*] as soon as [*indicate operation etc.*] is complete [*or as appropriate*]. In light of this explanation, perhaps you would be good enough to withdraw your notice requiring compliance within seven days.
>
> Yours faithfully
>
> Copy: Employer

16.3   Specific instructions

3.4   Apart from the architect's general authority to issue instructions, the contract contains only six other specific instances when the architect is empowered to issue instructions in particular circumstances:

(2.1.2)   ■ The architect may issue 'directions' to integrate the design of the CDP Works with the design of the Works as a whole. Although referred to as

'directions' the contract does not explain the difference between a direction and an instruction and this provision is made subject to another provision discussed above which refers to the architect requiring the contractor's consent for any 'instruction' affecting the design of the CDP Works.

■ The architect may instruct the contractor not to make good defects at its own cost (this provision is discussed in detail in Chapter 20).

■ The architect may instruct the ordering of variations by way of additions, omissions or other changes in the Works or the manner in which they are to be carried out.

■ The architect must issue instructions regarding the expenditure of any provisional sum (see Chapter 17).

■ The architect may issue instructions requiring the exclusion from the Works of any person employed there. The instruction is not to be issued unreasonably or vexatiously and is clearly intended to enable the architect to have incompetent operatives removed.

■ The architect must issue an instruction to confirm a change proposed by the contractor if the employer wishes to implement it. The instruction must refer to the change, any adjustment to the Contract Sum, the contractor's share of financial benefit and any adjustment to the date for completion. This is an unusual use for an instruction. The contract makes no reference to the adjustment to the completion date being made under the extension of time clause.

Any instructions given to the contractor's representative on the Works are deemed to have been issued to the contractor. In the context of the contract as a whole, such instructions must be in writing for the clause to have any effect, although in practice the architect seldom gives written instructions to the person in charge. Site instructions tend to be oral and are therefore ineffective unless or until confirmed, written instructions being sent to the contractor's main office.

16.4   Other instructions which will be empowered

It has already been mentioned that the courts will probably interpret the architect's general power under the contract to issue instructions quite narrowly. Although not specifically stated, it is considered that the following common situations will fall within the architect's powers:

■ *The correction of inconsistencies between the contract documents.* Although the contract deals with this matter, it does not specifically say that the architect can issue instructions. Nonetheless, it is considered that the architect certainly has that power: how else would the inconsistency be corrected?

■ *Opening up of work for inspection and testing of materials.* If the work or materials are found to be in accordance with the contract, the contractor will have a claim for extension of time. It is thought that, under those circumstances, the costs of carrying out the instruction could be valued, but if the contractor considers that it has been involved in loss and/or expense, that would have to be the subject of a common law claim (see Chapter 15).

- *The removal or correction of defective work.* The architect must have the power to issue these kinds of instructions.

- *Probably the postponement of any work in progress.* This power could not extend to deferring the giving of possession of the site which would amount to varying an express term of the contract – something which only the parties decide. Any postponement instruction would inevitably give rise to an extension of time. There would be no express entitlement to loss and/or expense under the contract and, because an authorised instruction cannot be a breach of contract, there would be no entitlement to damages for breach of contract either. The conclusion is that if the architect has power to postpone work as an implied term, there must be a further implied term entitling the contractor to further reimbursement. Certain categories of postponement concerning only a part of the Works could conceivably be brought under the existing terms: 'The Architect…may issue instructions requiring…change in the Works or the manner in which they are to be carried out…,' so enabling the architect to include direct loss and/or expense in the valuation.

3.6

135

3.6

16.5   Common problems

After failing to comply with a compliance notice, the contractor refuses to allow another contractor onto the site to carry out the instruction.

In that case, if the employer seeks an injunction from the court compelling the original contractor to allow access, it is likely that the injunction would be granted.

136

The behaviour of the person-in-charge towards the architect is disagreeable, aggressive and his speech is punctuated by obscenities. The architect has issued an instruction excluding him from the site. The contractor objects on the basis that he is an extremely competent operative and there is no one else to replace him.

The contract says that the architect may exclude any person from the site who is employed there. There is a proviso that the instruction must not be issued unreasonably or vexatiously. That means that, viewed objectively, it is reasonable for the architect to issue the instruction given all the circumstances. Moreover, the instruction must not be given maliciously or with the intention of annoying. One can easily imagine the architect becoming increasingly annoyed by the unacceptable behaviour of the person-in-charge to the extent that there is a strong temptation to 'put the dreadful man in his place'. The architect must put those feelings to one side and simply ask the question: 'Can I properly work with this person or is there a danger that the

3.8

137

project will be at risk as a result of his behaviour?' If the answer is that it is increasingly difficult to work with the person-in-charge and, particularly if the architect feels physically threatened by him, the architect will be justified in issuing the instruction.

The fact that the person-in-charge is a competent operative is irrelevant if the architect cannot work with him. The problem of replacing him lies with the contractor who should not have appointed him in the first instance.

17 Variations

17.1 Variations

Definition

The contract defines variation as:

- An addition to the Works;
- An omission from the Works;
- A change in the Works;
- A change in the order in which they are to be carried out;

3.6

- A change in the manner in which they are to be carried out;
- In the case of MWD, a change in the Employer's Requirements resulting in a change to the design of the CDP Works.

Additions and omissions

138

An addition to the Works is straightforward. An omission from the Works also sounds straightforward but it may not be so. The power to omit work means exactly that; the omission of the work. It does not entitle the employer to get another contractor to do the omitted work more cheaply. That would be a breach of contract, because the employer has agreed that the contractor is entitled to do all the work in the contract. Omitting work to give it to another contractor is not omitting work, it is taking the work from the contractor to give to someone else. This issue crops up quite often, particularly in domestic work, for which these contracts are generally used. The employer may suddenly realise that he or she can get some part of the work done by a friend at a very advantageous price. It is the architect's task to explain to the employer that if the work is omitted to give to a friend, the contractor will be able to make a common law (i.e. not under the contract) claim for damages for the breach of contract. The kind of damages which the contractor could claim is loss of profit on the work and/or materials item in the contract.

The JCT Minor Works Building Contracts 2016, Fifth Edition. David Chappell.
© 2018 John Wiley & Sons Ltd. Published 2018 by John Wiley & Sons Ltd.

Changes

A change in the Works is again straightforward and often includes elements of additional work and the omission of work as when the employer asks the architect to change concrete to brick paving laid in a decorative pattern.

Order of work

139

A change the order in which the Works are to be carried out is by no means clear. It is a general principle that a contractor is entitled to plan and perform the work as it pleases. The ability to plan the work in a specific way is an important power for the contractor because it enables it to plan economically when preparing its tender. Although the order of the work may be changed if an order is already stated in the specification or schedules, there is no power to create an order. Moreover, there is no provision for the architect to insert any dates against parts of the Works. Therefore, the contractor is under no obligation to complete any parts of the Works at any earlier time than the contract date for completion. Therefore, to summarise the position: the architect has no power to instruct the contractor to carry out the Works in a particular order. If the order of the Works is set out in the contract documents, the architect may instruct a change to that order, but the architect has no power to instruct that parts of the Works must be complete by any dates other than the overall contract completed date.

Manner of carrying out

140

The 2016 amendments to the contract have removed reference to a change in the period and substituted a change in the manner in which the Works are to be carried out. The precise meaning of that kind of change is not explained. The ordinary dictionary definition of 'manner' is 'the way a thing is done'. Therefore it seems that this gives the architect the power to instruct the contractor to carry out the Works in a certain way different from what the contractor intended. For example, the architect may instruct the contractor to excavate trenches using human labour, picks and shovels rather than by machinery. In giving the architect this kind of power, the contract seems to be allowing the architect to stray away from the architect's normal discipline and into contracting. Obviously if the architect instructs the contractor to do something in a particular way which is not what the contractor was intending to do, the architect will carry a large measure of responsibility for the success of the way in which the work is done. One would hope that if the architect has strong views about the way a particular activity should be done, the contractor would be canvassed for its views at an early date.

Employer's Requirements

A change in the Employer's Requirements resulting in a change to the design of the CDP Works seems clear enough but the architect must pay heed to the precise wording. The architect is not entitled to instruct a change to the design of the CDP Works, but only to instruct a change to the Employer's Requirements. If that instruction results in a change to the design of the CDP Works, it is a variation. A very important consideration, and one which the architect must keep firmly in mind when putting the contract documents together, is that the contractor's consent is required for this kind of variation. The contractor must be reasonable in delaying or withholding consent.

3.4.2

1.7

17.2 Valuation

The architect must value all the types of variation listed and also anything else which the contract says is to be treated as a variation. If unusually there is a quantity surveyor employed by the employer, the architect may take the quantity surveyor's advice about valuation. The contract is silent about the existence, identity or role of the quantity surveyor (see Chapter 7). Even if the architect does ask a quantity surveyor to undertake the valuation of variations, the architect bears the ultimate responsibility.

Agreeing a price

The architect must endeavour to agree a price with the contractor for any variation, but agreement must be reached before the contractor carries out the instruction. In practice, it allows the architect to invite and accept the contractor's quotation. On small works this is the sensible way of operating the provision if the additional or omitted work is anything other than more, or less, of the same.

3.6.2

Architect's valuation

If the architect does not agree a price with the contractor, the architect must carry out the valuation on a 'fair and reasonable basis'. Although it is pleasant to think that 'fair and reasonable' is an objective concept, in practice it is the architect's opinion as to what is fair and reasonable that matters. When carrying out the valuation, the architect must use, where relevant, prices in the relevant priced document. It is:

3.6.3

- The priced specification; *or*
- The priced work schedules; *or*
- The contractor's own schedule of rates.

It is a matter for the architect to decide whether the prices are relevant. It is considered that unless the work being valued is precisely the same, carried out under the same conditions, the architect is probably at liberty to ignore the prices in the priced document. This is because even a slight change in the conditions under which work is being carried out will have a marked effect on the cost to the contractor. The status of the contractor's schedule of rates has already been discussed (Chapter 1). If the architect waits until the first variation has to be valued before asking the contractor to provide it, the way is open for an unscrupulous contractor to massage the rates. Whether or not the contractor does so, the architect will retain suspicions. If the schedule is not provided before the job starts on site and calculations given to show that they are the same rates used by the contractor to produce its tender, any later provision of the schedule will be suspect and the architect will have an unenviable task trying to value variations properly.

Clearly a fair and reasonable valuation must also include the valuation of work or conditions not expressly covered in the instruction but affected by that instruction. For example, if the architect issues an instruction to change the doors from painted plywood faced to natural hardwood veneered and varnished, it might well affect the sequence in which the contractor hangs the doors, the degree of protection required and the difficulty of painting surrounding woodwork. All this must be taken into account in the valuation; it is 'direct loss and/or expense' associated with the variation, as is other disruption or prolongation directly flowing from the variation.

Information from the contractor

There is no express requirement for the contractor to submit vouchers or other information to assist the architect in arriving at a fair and reasonable valuation, but it would be a foolish contractor who refused reasonable requests in this respect. If the contractor refuses to supply information, the architect must carry out the valuation using reasonable endeavours and the contractor will have no valid claim if it receives less than it expects. The valuation will, of course, include an element for profit, overheads and so on, as usual.

Direct loss and/or expense

3.6.3

The valuation must include any direct loss and/or expense due to regular progress of the Works being affected by compliance with the variation instruction. The purpose of this provision is to reimburse the contractor for the loss and/or expense directly resulting from the variation, but not forming part of the cost of the varied work itself. In other words, it covers the effect of the introduction of the variation upon other unvaried work together with such things as scaffolding and, if the contract period is prolonged as a result, it also covers the extra site establishment costs (see Chapter 15).

17.3   Provisional sums

3.7

The architect must issue instructions to the contractor directing how provisional sums are to be spent. Provisional sums are usually included when the precise cost or extent of work is not known at the time of tender. The purpose is to have a sum of money to cover the cost. In domestic contracts provisional

3.6.3   sums are often inserted to cover the probable cost of things such as kitchen and bathroom fittings, and any other thing which the employer has not actually chosen at the time of tender. Provisional sums are not a good idea, because invariably what the employer chooses exceeds the allowance in the provisional sum and the Contract Sum is increased. When the architect issues the instruction, the provisional sum must be omitted and the instruction must be valued in accordance with normal valuation principles already discussed. It is possible to use a provisional sum to nominate a sub-contractor, but this is a very bad idea (see Chapter 9). Sometimes a contractor pricing the Works will put in its own provisional sum instead of a proper price. That kind of provisional sum has no validity and the architect should insist on a proper price being inserted. Only provisional sums put into the document by the architect have any validity.

17.4   Common problems

All the steelwork is CDP Works. The contractor has submitted its proposals for the design but the employer wants a change and engages a consultant structural engineer to redesign the steelwork and the architect issues an instruction to the contractor enclosing the structural engineer's design.

The architect is not entitled to do that. The contractor is responsible for the design, not the employer's structural engineer. The contractor could reject the instruction although, in practice, the contractor is unlikely to do so because if the contractor simply carries out the design, it can argue that it has no design responsibility for it. The contract is clear that what the architect should have done is simply issue an instruction to the contractor changing the Employer's Requirements so that the contractor retains full responsibility for redesigning the steelwork to suit the change.

If the contractor refuses to carry out a variation until its quotation has been accepted.

3.4 (3.4.1)
3.6.2

3.6.3

This is a very misguided action on the part of the contractor. The contract is clear that the contractor must comply with the architect's instruction forthwith. Although the contract states that architect and contractor must endeavour to agree a price prior to carrying out the instruction, it goes on to say that if there is no agreement, the architect must value the variation on a fair and reasonable basis using if relevant any prices in the priced document. The priced document will be either the specification or work schedules or a schedule of rates. There is little point having a contract with a document setting out the rates and prices if the contractor can simply insist on acceptance of a quotation before work progresses. The way the contract is intended to work is that the architect issues an instruction and then has a word with the contractor to find out what the contractor thinks the valuation should be. If the variation is simply omitting work which is in the contract to be done or if it is adding work which is the same as other work to be done, there should be no difficulty agreeing the valuation based on the priced document. Disagreement becomes more likely if the variation concerns work which is not similar to what is already priced. As soon as it becomes clear to the architect that agreement will not easily be reached, the contractor should be informed that the architect will carry out the valuation.

18 Payment

18.1 Important to read this first

The way in which the contractor is paid is similar, *but significantly different*, to the payment provisions in the previous editions of MW and MWD. The procedure is quite, some may say needlessly, complicated and what follows is as simple as it can be made without changing what the contract actually says. If you only read one chapter in this book – this is that chapter.

Purpose of the payment procedure

The purpose of the payment provisions is to set out the way in which the money is paid to the contractor. There is a series of interim payments in order to provide the contractor with cash flow, culminating in the final payment when all the Works in the contract have been completed and any defects rectified.

If the architect observes the procedure carefully, it will be the architect's certificates which determine the amount of money which is due to the contractor. However, if the architect fails to issue interim certificates in the correct form and at the correct time, the amount due may be determined by notices given by the contractor. This is particularly important if, as often happens, there is a dispute between architect and contractor about the value of the amount which should be paid. The interim payment procedure decides whether the employer or the contractor sits on the disputed amount while the actual amount payable is ironed out.

The interim valuation date – very important

Something called the 'Interim Valuation Date' is very significant under the 2011 MW and MWD contracts. Architects and contractors used to the date of issue of interim certificates being the important date for the start of the 14-day payment period must readjust to the new procedures. They are quite complicated but it is extremely important for all parties that they are properly operated. *The interim valuation date is the key date so far as payment is concerned.* The interim valuation dates under these contracts are the dates which trigger

The JCT Minor Works Building Contracts 2016, Fifth Edition. David Chappell.
© 2018 John Wiley & Sons Ltd. Published 2018 by John Wiley & Sons Ltd.

4.3

the payment process. An interim valuation date occurs every month starting with the first interim valuation date one month after the date for commencement stated in the contract particulars. Alternatively, the first interim valuation date may be inserted in the contract particulars.

Summary

The payment provisions in the contract are shown in the flowchart Figure 18.1 Briefly, they amount to this:

- There is an interim valuation date every month.
- There is a due date seven days after each interim valuation date.
- The architect must issue an interim certificate within five days of the due date and that is what the employer must pay.
- The contractor may submit an application to the architect no later than each interim valuation date stating the sum it considers to be due at the due date.
- If the architect fails to issue a valid interim certificate, the contractor may submit what is called a payment notice at any time and that is what the employer must pay, if the contractor has issued an application, that will automatically become the payment notice.
- In the case of an interim certificate, the final date for payment is 14 days after the due date.
- If the architect fails to issue the interim certificate correctly or at the correct time and a payment notice is submitted by the contractor after five days from the due date, the final date for payment is postponed by the same number of days as the notice was issued after the five-day period. Therefore, if the payment notice is issued two days after the five-day period, the final date for payment becomes 16, instead of 14, days after the due date.
- Whether the amount payable is decided by certificate or notice, the employer may issue a pay less notice (which replaces and which is somewhat different from what used to be called the withholding notice) not later than five days before the date for payment.
- Certificates and notices are to be issued even if the sum considered to be due is zero.

Read on for more details.

18.2   Contract Sum

4.1

The Contract Sum is the sum of money for which the contractor has agreed to carry out the whole of the Works (Article 2). It is stated to be exclusive of VAT which means that VAT payments which may be necessary will be additional to this sum. How far this will affect the employer will depend on the employer's status, from the point of view of being able to reclaim VAT, and the work involved in the contract. The architect must not include VAT in any certificate.

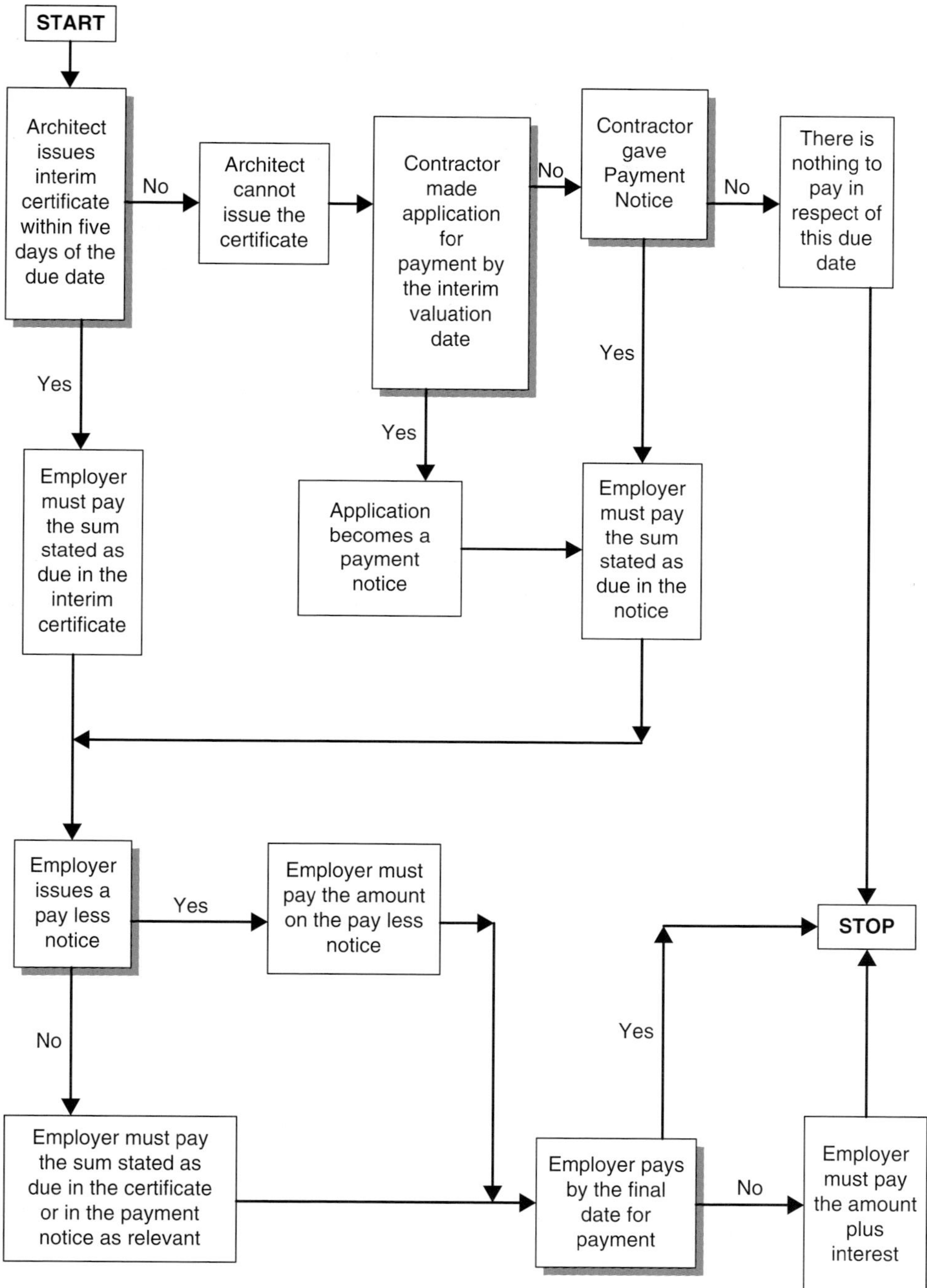

Figure 18.1   Flowchart showing interim payment provisions under MW and MWD.

The figure entered in the contract as the Contract Sum will be the contractor's tender figure or whatever figure the parties agree, probably after negotiation. The importance of the sum cannot be over-emphasised. MW and MWD are lump sum contracts which means that the contractor is entitled to payment

Table 18.1   Adjustment of Contract Sum under MW and MWD.

Clause	Adjustment
2.4 (2.5.1: MWD)	Inconsistencies in the contract documents
2.10 (2.11: MWD)	Defects etc. during the rectification period not made good
3.5	Non-compliance with instructions
3.6	Variations
3.7	Provisional sums
4.8	Computation of the final payment
4.3	Fluctuations
5.6.6	Making good of loss or damage
6.7.3	Consequences of termination under clauses 6.4 to 6.6
6.11.2	Consequences of termination under clauses 6.8 to 6.10

141   provided it completes substantially the whole of the Works. In theory, the existence of a system of interim payments does not alter the position and if the contractor abandons the work before completion, the employer is entitled to pay nothing more. Once written into the contract, the Contract Sum may only be adjusted in accordance with the contract provisions (see Table 18.1). Errors or omissions in the calculation of the Contract Sum are treated as accepted by employer and contractor unless, for example, it can be shown that the employer knew perfectly well that the contractor had omitted to take a page of the priced specification into account.

142

Error in calculation

Inconsistencies between documents must be corrected and if the correction results in an addition, omission or change in the Works it is to be treated as

2.4 (2.5.1)   a variation. (but see Chapter 1 which indicates that correction for inconsistency is unlikely). It is quite possible for the contractor to make a big error in its calculations to such an extent that the contract becomes no longer viable from its point of view. If the error is undetected before the contract is entered into, there is nothing the contractor can do about the situation, except perhaps submit an *ex-gratia* (on grounds of hardship) claim, with little hope of success. This situation must be avoided if possible because it is unsatisfactory from all points of view. The employer may indeed think that there is a financial advantage to be gained, but a contractor in this position has very little incentive to work efficiently and every reason to submit claims for additional payment at every opportunity. Unfortunately, unless the contract documentation includes quantified schedules or bills of quantities, it is very difficult to check the contractor's pricing. Some errors may be obvious, but where the contract documents consist of drawings and specification, the contractor's pricing strategy may be obscure unless it separately submits a detailed breakdown of the figure.

18.3   Interim certificates

4.3

The architect must issue an interim certificate no later than five days after each due date. The certificate must not be issued earlier than the due date or after the end of the five-day window. That is an unalterable rule. A certificate issued after the five days is simply invalid and of no effect. This cannot be stressed too much.

4.4.1

The contractor may, but need not, send monthly applications for the architect's consideration no later than the interim valuation date. The application must state the sum which the contractor believes is due on the due date and it must show how the sum has been calculated. However, it must be remembered that the certification does not depend upon a request from the contractor. Architects who are used to waiting until contractors submit applications for payment must change their methods or at some stage they will receive a nasty shock. More of that below.

In order to make the contract work properly, the architect must issue interim certificates to the employer of payments to the contractor. The employer cannot pay the certified amounts without knowing what they are. The employer must receive the certificate, while the contractor receives a copy. The interim certificate must state the amount the employer must pay and the basis on which the amount was calculated. The architect must attach to the certificate details of how the amount has been calculated. This ought to be a complete breakdown of the valuation showing each element of work and materials and the value of each element. Something less detailed than that may be acceptable provided that it shows how the architect has arrived at the amount in the certificate. Often, the architect will simply return the contractor's application with various amendments to show how the certified sum is calculated. The fact that a contractor may not make an application does not relieve the architect of responsibility for showing the calculation. If no application has been made, the architect may simply have recourse to the priced document, whether that is a priced specification, priced work schedules or a schedule of rates, in order to do the calculation.

The employer must pay within 14 days of the due date; not the date of issue of the interim certificate. If the interim valuation date is the 1 March, the due date will be seven days later on the 8 March. Therefore, the final date for payment will be 14 days later on the 22 March. If the architect takes a full five days after the due date to issue the interim certificate, i.e. on the 13 March, the employer will only have nine days from the issue of the interim certificate, not 14, in which to pay. In this example, even if the architect issued the interim certificate on the day after the due date, i.e. on the 9 March, the employer would only have 13 days in which to pay.

143

Although there is no provision for any other system of payment, it may be more convenient, on small works, to agree a system of stage payments. If it is desired to operate in this way, it is important to make the necessary amendments to the printed conditions and to ensure that the contractor is aware of the change at tender stage. It will have a considerable impact on its pricing strategy.

What must be included in an interim certificate?

The amount to be included in the certificate is to consist of:

- The percentage stated in the contract particulars (usually 95% before practical completion and 97.5% after) of the total value of work properly executed;
- Including amounts ascertained or agreed as variations or provisional sums or amounts ascertained following the contractor's valid suspension for non-payment;
- the value of any materials and goods which have been reasonably and properly brought on the site for the purpose of the Works, and which are adequately stored and protected against the weather and other things likely to cause damage;
- amounts (if any) calculated under the fluctuation provisions.

4.3.1

3.6

3.7

4.7

4.3.2

4.11

The contract states that the architect should state the total value of these items at the due date, but later states that they must be calculated as at the interim valuation date. There are seven days between the two dates and the amount of work properly executed and materials on site will no doubt increase between those dates. This appears to be a mistake. Pending any revision from the JCT, there are arguments for using either date. It is tentatively suggested that on balance the total value as at the interim valuation date should be used.

Amounts to be deducted must also be set out in each interim certificate as follows:

- The total amount stated as due in previous certificates;
- Any amount which has been paid by the employer in accordance with a payment notice given after the issue of the last certificate;
- Any deductions after instructing the contractor not to make good defects appearing during the rectification period;
- Any deductions following the contractor's failure to comply with an architect's instruction.

2.10 (2.11)

3.5

Each of the above items requires careful consideration, as follows.

The total value of work properly executed

The generally accepted view of 'value' is that it is the valuation obtained by means of reference to the priced document in relation to the amount of work actually carried out. Defective work is not included because it is not 'properly executed'. This contract does not specifically refer to the 'retention' but it is obviously the difference between the gross amount and the 95% (or whatever percentage is stated in the contract particulars) in the certificate. The retention is intended to deal with problems which might arise (see Chapter 20). The biggest problem would be if the contractor went into liquidation immediately following a payment.

'Properly executed' refers to the fact that the architect must not include the value of work which is defective, that is, not in accordance with the contract.

If the architect does certify defective work because, perhaps, the defect does not make itself immediately apparent, the correct procedure is to omit the value from the next certificate. This should pose no difficulties unless, of course, the contractor abandons the work first.

Amounts ascertained or agreed as variations or provisional sums

This refers to any variations which are valued before the date of the certificate and any instructions which the architect may issue with regard to provisional sums which result in an adjustment to the Contract Sum.

Amounts ascertained following valid suspension by the contractor for non-payment

If the employer has failed to pay an amount due to the contractor by the final date for payment and if the contractor suspends performance of its obligations under the contract after giving seven days' notice, the contractor is entitled to a reasonable amount for costs and expenses incurred by the suspension.

The value of materials and goods, etc.

The reasoning behind the inclusion of payments for unfixed materials on site is clear. It is to enable the contractor to recover, at the earliest possible time, money which it has already laid out. The architect need not include any materials which are considered to have been delivered to the site unreasonably early for the sole purpose of obtaining payment, but in its present form the clause contains serious dangers for the employer and for the architect who should not certify the value of materials or goods unless the contractor owns the materials or goods and has the right to pass ownership to the employer.

There is no provision for the contractor to provide proof of ownership before payment. Therefore, if unfixed materials are included in a certificate and the contractor does not own them, the employer could be faced with the prospect of paying twice for the same materials if the contractor goes into liquidation and the true owner claims the goods from site. Many suppliers and sub-contractors include special clauses which state that the contractor does not get ownership of the goods until the supplier or sub-contractor receives full payment. These kind of clauses are called retention of title clauses. The prevalence of retention of title clauses in the supply contracts of builders' merchants makes this a very real danger. The general law states that when materials and goods are incorporated into the building, they become part of the land and cease to be independent things. This will usually defeat a retention of title clause.

The position is actually quite complicated but, briefly, while the materials remain on site as materials, the supplier may reclaim them if the contractor becomes insolvent and cannot pay. However, if the materials have been incorporated into the building, the supplier may be prevented from recovering them because, they have become part of the employer's property. In any event the supplier has no rights to enter upon the employer's property and even if the supplier was able to enter, the goods would have to be removable without damage to the remaining construction.

4.3.2

Some architects amend the contract by deleting the whole of the part about payment for materials. Since there is then no provision for payment for off-site materials, this will mean that the architect does not have to certify any unfixed materials whatsoever. However, it will certainly influence the contractor's tender price.

Fluctuation provisions

4.9,
Schedule 2

Most MW and MWD contracts are stipulated to be fixed price but there is provision for fluctuations in contribution, levy and tax to be taken into account if desired.

The total amount due in previous certificates

This is hopefully self-explanatory. However, it is sometimes confused with what has been previously paid. In the ordinary course of events, the architect will not necessarily know what has been paid. All the architect will know is what has been previously certified and it is this figure which must be deducted. If the employer has failed to pay one or more certificates, the contractor will often appeal to the architect for assistance. But other than advising the employer that certificates ought to be paid promptly, there is nothing that the architect can do and architects are well advised to leave this kind of dispute for the employer and the contractor to sort out between themselves. It is not the architect's responsibility.

Sums paid in accordance with a payment notice

4.4.2.2

4.3

4.4.2.1

A payment notice may be given to the employer by the contractor if the architect fails to issue an interim certificate within five days after the due date or if the certificate is not issued properly: for example, if the certificate does not state the sum due from the employer or does not state or demonstrate the basis of calculation of the sum. A payment notice may not be given by the contractor under any other circumstances. A payment notice must itself state the sum due on the due date and must state or demonstrate the basis of calculation. If the contractor has issued an application for payment no later than the interim valuation date, that will automatically become the payment notice. Given what was said in the last paragraph, it is strange that the contract requires the architect to deduct what the employer has *paid* in response to a payment notice given since the last certificate. Nevertheless, that is what the architect has to do if the contractor properly gives a payment notice.

4.5.3

Importantly, the final date for payment is postponed by the same number of days as the number of days after the expiry of the five days in which the architect could have issued a certificate. What that means is best explained by an example. If the due date is the 5th of a month, the final date for payment would be $5 + 14 =$ the 19th of the month. If the architect failed to issue an interim certificate within five days of the due date, the contractor might issue a payment notice two days later. In that case the final date for payment would be postponed for two days until the 21st of the month.

Pay less notice

4.5.4.1

An employer who does not intend to pay the amount stated in an interim certificate or in a payment notice given by the contractor may issue a pay less notice no later than five days before the final date for payment. It is important that a pay less notice must state the sum which the employer thinks is due to the contractor on the date at which the notice is given. It must also state the basis of calculation. It is important that all these little details are strictly observed or the contractor may be able to successfully challenge the validity of the notice on a technical point. The architect or any other person nominated by the employer

4.5.5

may give the pay less notice.

18.4 Final certificate

The contract lays down a precise time sequence for the events leading up to, and the issue of, the final certificate (see the contract time chart, Figure 18.2).

4.8.1

The contractor's duty is to send the architect all the documentation the architect reasonably requires in order to calculate the amount to be finally certified. The architect is probably entitled to request any particular supporting evidence necessary. The contractor has three months, or whatever other period is inserted in the contract particulars, from the date of practical completion to send the information to the architect.

2.10 (2.11)

4.8.1

2.10 (2.11)

Although the contract does not say so, it is clear that this three-month period is related to the three-month rectification period, so that if the rectification period is increased to, say, 12 months, the period allowed for the contractor to provide all the information for the calculation of the final payment must be the same. This is because the architect has power to issue instructions about the making good of defects. It may be that in some instances it involves an adjustment to the Contract Sum, in which case the amount must be reflected in documents to be provided by the contractor. In those circumstances, different time periods are not workable.

2.11 (2.12)

The last due date is 28 days after the receipt of the contractor's information or after the certificate of making good is issued whichever is last. In practice, the certificate of making good will often be the deciding factor. The final certificate cannot be issued until after the due date. The architect then has just five days to issue it. If the contractor is late in sending the information to the architect, the contractor is technically in breach of contract, but it is of little consequence. The contractor is simply delaying the time when it receives payment

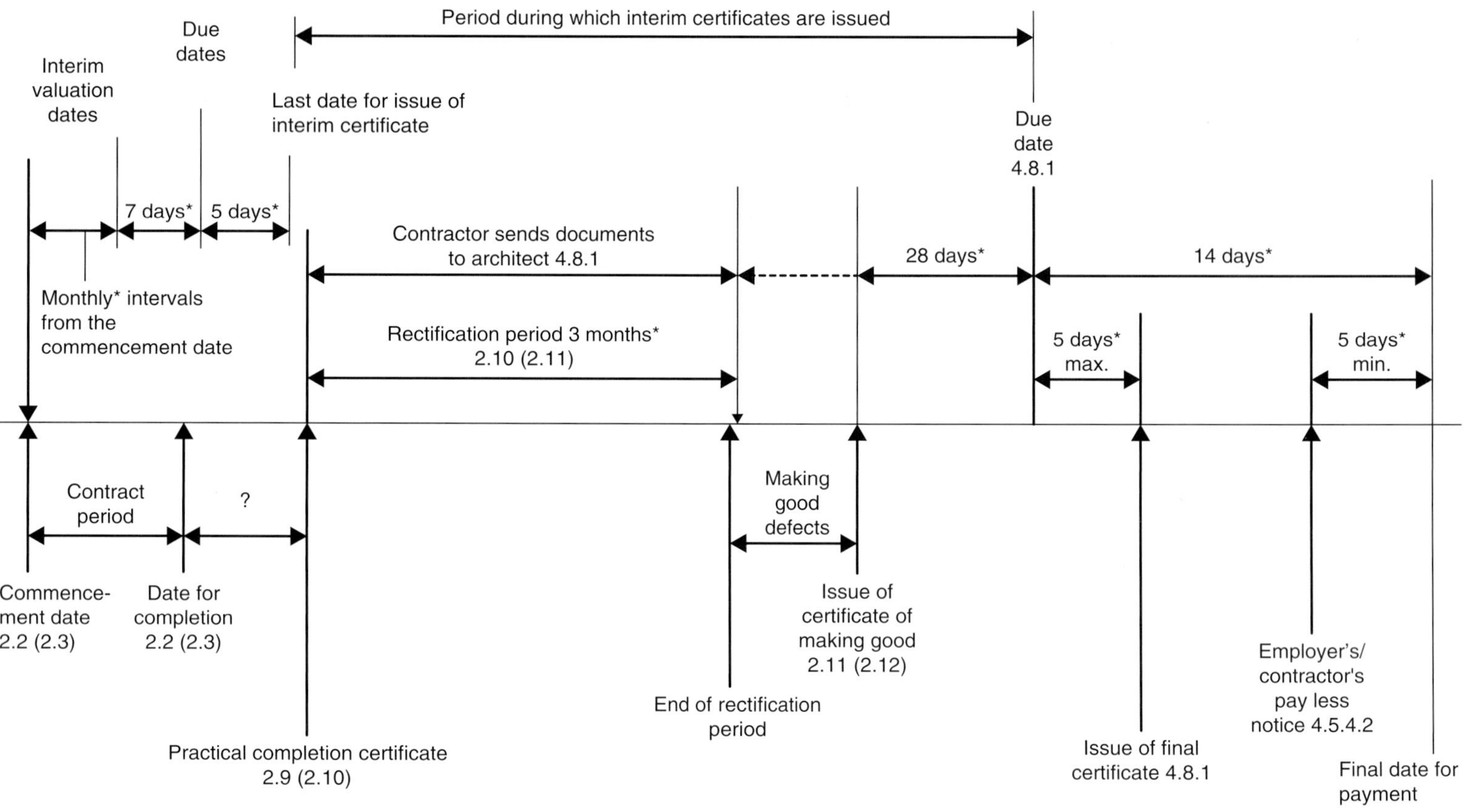

Figure 18.2 MW and MWD time chart.

because the 28 days does not begin to run until the latest of receipt of the contractor's documentation or the issue of the certificate of making good.

Having said that, the contractor cannot delay the due date indefinitely by failure to supply the information needed for the calculation of the final certificate. Once the certificate of making good has been issued, it is suggested that lack of information from the contractor must not delay the issue of the final certificate. Although it is customary for the contractor to submit its version of the final account during this period, it is clear from the contract that it is the architect who must compute the final account. If the contractor has failed to supply the information by the time the architect has issued the certificate of making good and overall is unreasonably late, it is suggested that the architect should write to the contractor. The letter should inform the contractor that if the architect does not receive the information by (a stated date), the 28 days to the due date will start to run from the stated date as being the latest date by which the Contractor ought reasonably to have provided the information.

Throughout the contract period, the architect should have been keeping a running total of the prospective final account and obtaining documents from the contractor where necessary to carry out the valuations. The period after practical completion is to allow the contractor to draw the architect's attention to any further information that is relevant. It is not intended that the contractor submits nothing until this point and then submits vast bundles of documents. If that is what had been intended, 28 days would have been an entirely inadequate period for the architect to prepare the final account.

The final certificate must state the amount remaining due to the contractor or, more unusually, the amount due to the employer. The latter situation will only occur if the architect has previously over-certified. The certificate should state to what the amount relates and how it has been calculated. Although not expressly required under the contract, it is good practice for the architect to provide the contractor with a final account before issuing the final certificate and reference to this should be sufficient. The employer or the contractor as the case may be has 14 days from the due date, as before, in which to pay. There is no provision in the contract for the contractor to agree the computations before the final certificate is issued, but it is customary to attempt to obtain the contractor's agreement, because the final certificate under this contract is not conclusive about the amount certified. If the contractor delays sending agreement to the architect this does not affect the architect's duty to issue the final certificate within five days of the due date and there is no method by which this date can be prolonged. A letter from the architect to the contractor should make this clear (Box 18.1). The sum stated in the final certificate as payable to the contractor or the employer is the sum to be paid subject only to any pay less notice.

Payment notice

If the architect fails to issue the final certificate in proper form, stating the basis of calculation of the sum due within five days of the due date, the architect's power to issue the certificate is gone. Then the contractor may issue a payment

> **Box 18.1   Letter from architect to contractor requesting agreement to the computation of the final sum.**
>
> Dear Sirs
>
> PROJECT TITLE
>
> I enclose two copies of the computation of the final sum to be certified as payable to the contractor/employer [*delete as appropriate*].
>
> I should be pleased if you would signify your agreement to the sum and the way it has been calculated by signing and dating one copy of the calculation in the space provided and returning it to me by the [*insert date*] at the latest.
>
> If you have any queries, please telephone me as soon as possible, but you should note that, in any event, I have a duty under clause 4.8.2 of the conditions of contract to issue my final certificate no later than [*insert date*].
>
> Yours faithfully
>
> Copy: Employer

4.4.2.2 *4.5.4.1*	notice stating the sum due on the due date and must state or demonstrate the basis of calculation. The sum so stated becomes the amount payable subject only to the employer's pay less notice. It is extremely unlikely that the contractor will issue a payment notice unless it believes that there is money due to it. Unsurprisingly, the contract does not provide for the situation where a contractor issues a payment notice in favour of the employer.

Pay less notice

4.5.4 *4.5.5* *4.5.4*	An employer or a contractor who does not intend to pay the amount stated in the final certificate may issue a pay less notice no later than five days before the final date for payment. It is important that a pay less notice must state the sum which the employer or the contractor thinks is due to the contractor or the employer (as the case may be) on the date at which the notice is given. It must also state the basis of calculation. If these details are not strictly observed, the validity of the notice may be successfully challenged on a technical point. The architect or any other person nominated by the employer may give the pay less notice on behalf of the employer. The sum stated in the pay less notice is the minimum sum payable to either the contractor or the employer as appropriate. The employer may issue a pay less notice in respect of a payment notice issued by the contractor provided it is issued no later than five days before the final date for payment.

18.5   Effect of certificate

It is refreshing to note that no certificate is stated to be conclusive. Thus, the issue of the certificate of making good does not preclude the employer from asserting that the contractor is liable for anything which is not in accordance with the contract. Similarly, if the architect inadvertently includes defective work in an interim certificate, the situation can be remedied with the next certificate issued.

The contractor's liability is not reduced in any way by the issue of the final certificate which is not even conclusive as far as the calculations are concerned.

18.6   Failure to pay

If the employer fails to pay any amount due to the contractor under any interim or the final certificate or any payment notice by the final date for payment, the employer must pay simple interest on the outstanding amount at the rate of 5% over the dealing rate of the Bank of England current on the date payment becomes overdue. This is a substantial amount although not as substantial as the current statutory percentage which requires 8% above Base Rate together with a lump sum payment and possibly other costs incurred.

1.1, 4.6

149

If the final certificate shows a payment due from the contractor to the employer and the contractor fails to pay by the final date for payment, the contractor must pay simple interest to the employer at the same rate.

4.6

The acceptance of interest does not imply the relinquishing of any right to the actual payment itself. Moreover, acceptance of interest on unpaid money in interim certificates does not prevent the contractor from suspending performance of its obligations (see Chapter 5, section 5.8) or indeed from terminating its employment.

4.7

6.8

18.7   Retention

There is no reference to retention by name in this contract, but only by implication. The contract only refers to the 'applicable percentage' to be certified. There is no reference to:

4.3

- The purposes for which the retention can be used;
- Its status as trust money;
- Keeping it in a separate bank account.

The reason for this is clearly to maintain the brevity of the contract as a whole, but there may be repercussions:

Use of retention
It used to be allowable for the employer to deduct the cost of employing others to carry out instructions with which the contractor had failed to comply, from

3.5

monies due to the contractor. Retention monies clearly fell within this description, but now the cost is to be deducted from the Contract Sum. If the contractor goes into liquidation before the Works are complete, there is no doubt that the employer would be entitled to make use of retention monies to complete the work. There is no express provision enabling the employer to make use of the retention for any other purpose. For example, the employer may not deduct the cost of taking out insurance if the contractor defaults. In practice, as mentioned earlier, the situation may suggest taking a broader view.

Trust money

Unlike the corresponding provisions of SBC, IC and ICD, the retention is not expressly stated to be trust money and it is not considered that any term would be implied from the general law to create such a trust.

Whether the retention is or is not a trust is of vital importance to the contractor for two reasons. First, a trust is governed by statute and, despite what the contract provisions may state, it is possible that the employer always has a duty to invest and to return any interest to the contractor. Where the retention is not stated to be a trust, the contractor is not entitled to any interest.

Second, where the retention is stated to be trust money, the employer is in the position of trustee, merely holding the money for the contractor's ultimate benefit. It is, in no sense, the employer's money. Therefore, if the employer was to become insolvent, the trust money would not form part of the employer's assets in the hands of the trustee in bankruptcy or the liquidator. The contractor would be able to recover the full amount provided that it was clearly separated from the employer's other money and held in a separate account. If the money is not a trust, as in the case of MW and MWD, the contractor would have no better chance of recovery than any other unsecured creditor. MW and MWD, therefore, leave the contractor at a severe disadvantage.

18.8 Common problems

The employer fails to issue a pay less notice but there are many defects in the Works. Does the employer have to pay?

The general answer is that the employer must pay the amount in the interim certificate or in the payment notice, whichever is relevant. That is because the interim payment provisions are not intended to finally decide whether the employer or the contractor gets the money. They are merely to decide who holds the money until the final certificate is issued. The provisions actually favour the employer, because the employer, or the architect on behalf of the employer, has only to issue a pay less notice in the proper form and at the right time to retain the money. Therefore, if the pay less notice is not properly issued, the employer must pay. However, there is one exception. If the contractor becomes insolvent before payment is made, even if the employer has wrongly held on to the money and the final date for payment has long gone, the employer can retain the money, at least until the concluding accounts after termination are drawn up.

150

Sixth months after the issue of the final certificate, the contractor says that the final certificate was undervalued and unless the employer makes payment of the shortfall, the contractor will claim the amount in adjudication.

Under IC, ICD and SBC contracts the final certificate is conclusive evidence (that is to say that it cannot be challenged) in any proceedings about a number of things. One is that, unless the certificate is challenged within a stated period, all the clauses in the contract stating how the Contract Sum must be adjusted have been correctly applied. What that amounts to is that the figure in the final certificate is not subject to challenge. A final certificate issued under MW or MWD is different. It is not conclusive evidence about anything. It is simply the architect's professional opinion about the amount due to the contractor. Therefore, the contractor (and of course the employer) can challenge the figure at any time to the end of the limitation period (six years if the contract is under hand and 12 years if a deed).

151

19 Practical completion

19.1 Practical completion

Definition

Despite the significant consequences of the architect's certificate, 'practical completion' is nowhere defined in the contract. It does not mean substantially or almost complete and the precise meaning has exercised several judicial minds. It means complete for all practical purposes. On balance, it seems to mean the stage at which there are no defects apparent and only very trifling items remain outstanding. What qualifies as 'trifling' will depend on circumstances. It is not thought that the architect is justified in withholding the certificate until every last screw and spot of paint is in place. That would indeed be completion, but the contract clearly intends something rather short of that by the use of the word 'practical'. Within these guidelines the architect is free to exercise some discretion. The architect would not be justified in issuing a certificate, in any event, if items remained to be finished which would seriously inconvenience the employer.

Many architects issue certificates of practical completion with a list of defects attached and sometimes with a qualification that the certificate is 'subject' to the list of defects being rectified. That is not acceptable. The contract does not allow a qualified certificate and a certificate that states anything other than the date of practical completion or if it has a list of defects attached, it is probably invalid.

19.2 The contract says

2.2 (2.3)

The contract states that the Works must 'be completed by' the date to be inserted on the contract particulars. The words have their ordinary meaning, that is to say the completion of the Works must not take place after the stated date, but it may take place before the stated date.

The JCT Minor Works Building Contracts 2016, Fifth Edition. David Chappell.
© 2018 John Wiley & Sons Ltd. Published 2018 by John Wiley & Sons Ltd.

No partial possession

There is no provision for partial possession by the employer. This omission is entirely reasonable in view of the small-scale nature of the Works for which this contract is intended to be used and the correspondingly short time-scale. It is extremely unlikely that the employer will need to take possession of part of the Works before completion. Occupation by the employer is not equivalent to practical completion. If it is decided that provision must be made for partial possession it is worth considering the use of IC or ICD instead. If phased completion is to be provided, MW and MWD are not suitable and either SBC, IC, ICD or ACA 3 would be appropriate. It is not envisaged that partial possession, much less sectional completion, would be a feature of work for which MW or MWD were being considered.

155

Certification

The architect must certify the date when in his or her opinion the Works have reached practical completion. In addition, the contractor must have complied adequately with any obligations in the CDM Regulations to supply documents or information. If MWD is being used there is an additional requirement to supply documents and information necessary to explain the CDP. Therefore, the architect is to certify the date by which all criteria have been satisfied. Only one date may be certified and not, as some architects believe, one date for each criterion. The architect must issue the certificate in writing. Although no time scale is indicated, the architect must issue the certificate within a reasonable time of the certified date because it is a particularly important stage in the contract (see below). In practice, the certificate should be issued immediately.

2.9 (2.10)

3.9

(2.1.3)

1.6.1

Note that it is the architect's opinion which is required by the contract, not that of the employer or the contractor. Note that the architect cannot issue the certificate even when in the architect's opinion physical practical completion has been achieved. The CDM provisions must also be satisfied (see above).

The architect's duty to issue a practical completion certificate does not depend on any request by the contractor. The duty must be carried out as soon as the architect is satisfied that the criteria have been satisfied. Many architects arrange a hand-over meeting to which the employer and any consultants are invited. It should be remembered, however, that the architect cannot transfer responsibility for certifying practical completion to the employer, although it obviously prudent to see that the employer is happy with the building before taking possession. In practice, it is much more likely that the employer will wish to take possession before the architect is thoroughly satisfied.

In such circumstances the architect must strictly observe his or her duty and refuse to issue the certificate until satisfied that practical completion in all its aspects has been achieved. The contractor will then gain no benefit and may be at some disadvantage in completing the work. The contractor may complain to the employer who may, in turn, complain to the architect who should put the position to the employer in writing for the record (Box 19.1). If the architect

> **Box 19.1   Letter from architect to employer if possession of the Works has been taken before practical completion.**
>
> Dear Sir
>
> PROJECT TITLE
>
> I refer to your letter of the [*insert date*].
>
> In my opinion, practical completion has not been achieved. Therefore, I have no power to issue a certificate. I note that you have agreed with the contractor to take possession of the building. My advice is that you are unwise to do so. Taking possession before practical completion, even with the contractor's agreement, may lead to problems. For example, the insurance position will not be clear and you should seek advice from your broker. In addition, the contractor may seek additional payment on the grounds of having to work in an occupied building. I will continue to inspect until practical completion has been achieved. At that date the rectification period will commence. Naturally, the contractor has a great interest in obtaining the certificate of practical completion and he may well complain until I feel able to issue my certificate.
>
> Yours faithfully

submits to pressure, it may well leave open a future claim for negligence, not only from the employer but also from third parties to whom the architect may have given a warranty that duties will be carried out with reasonable skill and care.

Pre practical completion 'snagging'

The architect is under no obligation to issue lists of outstanding items if the certificate is withheld. Clerks of works often consider it part of their duty to supply the contractor with so-called 'snagging lists'. It is a bad practice because the contractor tends to assume that when it has completed the lists, its obligations are at an end; therefore, disputes sometimes occur. The contractor's obligations should be clear from the contract documents and the duty to make sure that the work is complete in accordance with the contract lies with the contractor, not with the architect or the clerk of works (if employed). More particularly, any 'snagging lists' should be prepared by the contractor's person in charge as part of the normal supervision of the Works. Whether that is done is not the architect's direct concern.

A word about 'snagging': the contract does not refer to 'snagging' anywhere. The contract refers to defects. Architects and contractors often refer to 'snags' and 'snagging' as though 'snags' are something different and somewhat less important than defects. The fact is that whether one refers to defects or snags, one is referring to work or materials which is not in accordance with what the contract says the contractor should do. In legal parlance defects and snags are breaches of contract for which the contractor is liable. The contractor is not less liable if a defect is referred to as a 'snag'. In order to avoid misunderstandings, 'snags' and 'snagging' should not be used, certainly not by the architect who should know better. The words 'defects' and 'inspections for defects' should be used instead.

19.3   Consequences of practical completion

The issue of the certificate of practical completion is of particular importance to the contractor because it marks the date when:

2.10 (2.11)   ■   The rectification period commences;

2.8.1 (2.9.1)   ■   The contractor's liability for liquidated damages ends;

■   The employer's right to deduct full retention ends and half the retention

4.3   held becomes due for release;

■   The machinery culminating in the issue of the final certificate is set in motion and the contractor must supply all the documents reasonably

4.8.1   required for the calculation of the final payment;

5.4A   ■   The contractor's liability to insure ends.

19.4   Common problems

The Works were completed three months ago but the contractor has only now provided all the information required by the CDM Regulations, therefore the architect has issued a certificate of practical completion but back-dated it three months.

The architect is not entitled to do that. Practical completion can only be certified when the Works have reached practical completion and the contractor has

2.9 (2.10)   provided the CDM information. Therefore, the date of practical completion is when both of those conditions have been satisfied. In addition, if MWD is being used, all the drawings and specifications must have been provided to the architect in respect of the CDP work. To back-date the certificate three months would be to certify practical completion before it had occurred resulting in an invalid certificate which the employer could challenge.

Must the architect issue a practical completion certificate if the contractor's employment has been terminated but the Works have been completed by another contractor?

No. The original contract is no longer operative and the new contractor will have been engaged on a different contract. The architect will issue a practical comple-

6.7, 6.11   tion certificate under the new contract, but the only things that can be done in relation to the old contract are the things set out in the termination provisions.

Very often, if termination has been the result of the contractor's insolvency, the receiver or liquidator will press the architect for a practical completion certificate and payments, because the receiver's and liquidator's object is to collect as much money as possible as quickly as possible. Surveyors are often employed to try to frighten architects into making hasty and often unenforceable decisions. The contract procedures must be allowed to take their own pace.

The architect has agreed with the contractor that a practical completion certificate will be issued even though there are still defects outstanding and the contractor has promised that he will rectify the defects and complete all outstanding work within the next three weeks.

156

This is not a good idea on many levels. The architect has no power to issue a practical completion certificate if there are visible defects. Therefore, this certificate is invalid and of no effect. The contractor's promise is worthless and unenforceable, because he is simply promising to do things which he is already contracted to do. There is no evidence of a new contract between employer and contractor. It cannot be argued that the early issue is a change for the benefit of the contractor because, as noted above, the certificate is invalid. Therefore, the architect's insistence that the contractor must comply with its promise is misplaced. In this situation, the employer must make clear that both the certificate and the contractor's promise are worthless. Therefore, practical completion has not yet occurred, and the contractor must rectify all defects and get the Works to a stage of practical completion. Until then, liquidated damages will continue to run. If architect and contractor do not accept this analysis, it may require the employer to seek a declaration from an adjudicator that the certificate is invalid.

20 Defects liability

20.1 During construction

2.1

There is nothing in the contract which expressly deals with defects during construction. The contract requires the contractor to carry out the Works in a proper and workmanlike manner and in compliance with the contract documents. Therefore, it is clear that the contractor must not carry out the Works in any other way. Indeed, to carry out the Works in any other way produces defects. Put another way; a defect is something which is not in accordance with what the contract requires but, unlike the position in other contracts, MW and MWD do not refer to this until after practical completion. Obviously some defects may arise because the architect has not designed or specified correctly and the contractor is not liable for those. Some architects think that the contractor has the responsibility of making sure that the architect's design and specification work properly. In some limited circumstances, the contractor may have a duty to warn of errors in the architect's design, but the instances are likely to be few and concerning serious defects.

157

3.4

So what is the position if the architect finds defects in the Works before practical completion? The architect's power to issue instructions must extend to the power to instruct the contractor to rectify defective work or materials. If the contractor fails to do so, the architect can issue a written notice giving the contractor seven days in which to comply. If the contractor fails to comply, the employer may get another contractor to do whatever is necessary to comply and the architect may deduct the costs from the Contract Sum.

3.5

20.2 During the rectification period

It is not often realised that the rectification period is inserted in the contract for the benefit of both parties. It allows a period of time for defects to appear and to be corrected with the minimum of fuss and cost. Any defect which is the fault of the contractor is a breach of contract on the part of the contractor and, without such a period, the employer would have no contractual remedy. The employer would be left to the common law right to seek damages. Moreover, and more importantly, if there was no rectification period, the contractor would have no right to re-enter the site to remedy the defects. If the employer suffers some loss as a direct result of the defects and the mere remedying of those defects is not adequate restitution, action at common law is always available to obtain damages from the contractor for the breach.

Contractors and employers commonly hold two mistaken views about the rectification period:

1. That the contractor's liability for defects ends at the end of the rectification period.
2. That the contractor is liable for anything which is showing signs of distress or with which the employer is not satisfied.

The contractor's liability does not end at the end of the rectification period. What does end is its privilege to return and remedy the breach of contract. After that time, the contractor remains liable for defects until the expiry of the limitation period (see Chapter 3) and, after proper notice, the employer can simply pursue an action at common law for damages. Obviously the employer must mitigate the loss. That is to say that the employer must do whatever can reasonably be done to reduce the amount of loss claimed from the contractor. This may best be done by inviting the contractor to return to rectify the breach.

The second mistaken view probably owes its origin to the practice of referring to the rectification period as the 'maintenance period'. Architects and contractors alike are guilty in this respect (even GC/Works/1 (1998) and ACA 3 use the term). Maintenance implies a heavier responsibility than simply making good defects. Re-polishing, cleaning and generally keeping a building in pristine condition is maintenance, but the contractor has no responsibility for that. The employer's dissatisfaction with the building is also of little consequence in itself if the contractor has carried out its obligations. If instructed to carry out what amounts to matters of routine maintenance, the contractor should write to the architect making the position clear (see Box 20.1).

20.3   Defects, shrinkages and other faults

2.10 (2.11)

158

The contract requires the contractor to make good defects, shrinkages and other faults. The phrase 'other faults' seems to add little if anything to the contractor's liability, because it means others faults like defects and shrinkages. A qualification is that the defect must be due to materials, goods or workmanship not in accordance with the contract. If the employer is unhappy about the paintwork, it could be that the contractor has not applied it correctly in accordance with the specification. On the other hand, it could be that the specification is inadequate. Only the former situation would give rise to liability on the part of the contractor. An inadequate specification is usually the architect's responsibility. If MWD is used, defects due to the contractor failing to comply with its CDP duties must also be made good. Obviously, under MWD the contractor will have both a design and construct liability for some parts of the Works.

Box 20.1   Letter from contractor to architect regarding routine maintenance.

Dear Sir

PROJECT TITLE

We have received your instructions dated [*insert date*] which you purport to issue under the provisions of clause 2.10 [*substitute '2.11' when using MWD*] of the conditions of contract. Among the items listed as defects for us to make good are [*list defects complained of*].

Clause 2.10 [*substitute '2.11' when using MWD*] requires us to make good defects, shrinkages or other faults provided that, among other things, they are due to materials and workmanship not in accordance with the contract. What you are in effect instructing us to do is to carry out items of routine maintenance. That is not our obligation under the contract and you have no power to issue such an instruction.

Yours faithfully

At one time, the contract used to refer to 'excessive' shrinkages. The intention appeared to be to exclude those shrinkages which could be said to be an unavoidable consequence of building operations. This was an eminently sensible approach in principle, but one which could result in a dispute because what was excessive to the architect may have been trifling to the contractor. Indeed, the whole question of shrinkages is fraught with difficulty. They are the contractor's liability only if they result from workmanship or materials which are not in accordance with the contract. In practice, since the employer holds the purse-strings, it is the contractor who has to convince the architect that the shrinkage is not its responsibility. Shrinkage usually occurs due to loss of moisture or thermal movement. A common example is the shrinkage which occurs in timber after the building is heated. The architect will have specified a maximum moisture content for the timber, and subsequent shrinkage can only be because the timber was installed with, or allowed to develop, too high a moisture content, or the architect's specification was wrong or the employer has maintained the building at an unreasonably high temperature.

It is no defence for the contractor to say that the architect's specified moisture content was impossible to achieve under normal site conditions. It is well known that it is difficult to maintain a low moisture content under the normally damp conditions which prevail on site, but that is not to say that such conditions cannot be improved by the use of suitable temporary heaters, ventilation, etc. The contractor's obligation is to provide workmanship and materials in accordance with the contract and it would have been well aware of the architect's requirements at tender stage. What it is really saying, therefore, is that it found it too expensive to comply with the architect's specification, and that is no defence at all.

The 'excessive' qualification to 'shrinkages' has been dropped for the MW and MWD contracts and, therefore, all shrinkages are now included provided only that they result from materials and workmanship not being in accordance with the contract.

20.4   Frost

The contractor's liability to make good frost damage is no longer expressly stated but in any event it would be limited to damage caused by frost which occurred before practical completion. This is perfectly reasonable since the contractor was in control of the Works up to, but not after, practical completion. Damage due to frost occurring after practical completion is the responsibility of the employer. In practice, there should be no great difficulty in detecting the difference. Frost damage after practical completion may be due, for example, to faulty detailing, unsuitable materials or lack of proper care by the employer. Note that the test is not when the damage occurred, but when the frost, which resulted in the damage, occurred.

20.5   Procedure

There is a process to observe if the defects which appear during the period are to be dealt with appropriately. It is surprising how often problems arise. This is partly due to the tendency of many contractors to walk away from both defects and retention to concentrate on other work.

Length of period

The rectification period starts on the day after the date of practical completion stated in the architect's certificate. If no period of time is inserted in the contract particulars, the period will be three months. There is no reason to limit the period to three months, because there is no connection between the length of the contract period and the rectification period. If the period is written into the contract as longer than three months, the period in the contract particulars for the contractor to supply documents for the calculation of the final payment should also be extended by the same amount. That is because the discovery and notification of defects may sometimes require additional work not strictly due to the defect. Whether the contract period is long or short, there will be some items of work completed just before practical completion and it is these items which the architect must consider when advising the employer of the length of rectification period required in a particular case.

In general, there is probably much to be said for always inserting twelve months as the length of period on the basis that the building will be tested against all four seasons of the year. It is probably true that the contractor will include a slightly increased tender figure if twelve months is included as the rectification period, but there is really no reason why it should do so. It probably stems from a mistaken idea of the limits of its liability (see section 20.2). The final certificate will, of course, be delayed, but only 2½% of the total value of the Works will be outstanding.

Appearance of defects

The defects etc. which the contractor is to make good are those which 'appear within' the rectification period. The wording suggests that any defects which have already appeared before practical completion could not be included as defects which the contractor must make good. In practice, the situation is not as bad as that. Defects which were apparent (sometimes referred to as 'patent defects') before practical completion would preclude the architect from issuing the certificate of practical completion (see Chapter 19). Moreover, since no certificate is conclusive under this contract, the contractor's obligation to carry out the work in accordance with the contract is not reduced by the issue of any certificate and if the contractor is so misguided as to refuse the opportunity of remedying the defects during the rectification period, the employer retains the normal common law rights intact.

The danger is that if the architect certifies practical completion while there are some visible defects, only a small amount of money will be retained and the contractor may never return. The employer's common law rights will be useless if the contractor has gone into liquidation and the employer may say that the architect was negligent in the issue of the certificate and look to the architect for recompense. If the architect does overlook some defects before issuing the certificate, the sensible thing to do is to include them and any work not completed with the defects which actually appear 'within' the period. This may not be precisely what the contract says, but it does no violence to the contractor's rights.

159

Notification

The architect has until 14 days after the end of the rectification period to notify the contractor in writing about the defects. Notice is essential before the contractor can become liable. It is prudent to organise an inspection a few days before the period expires so that the architect can issue the final list of defects on the final day. The contractor should be prompt to respond if some items are not its responsibility (see Box 20.2). If there is an urgent defect, such as a burst pipe or leaking roof, the architect may notify the contractor about such defect before the end of the period. In any event, it is likely that the architect can use the powers in the contract to instruct the contractor to carry out urgent remedial action during the rectification period. All defects are to be made good by the contractor entirely at its own cost unless the architect instructs otherwise.

160

3.4

Box 20.2   Letter from contractor to architect after receipt of schedule of defects.

Dear Sir

PROJECT TITLE

Thank you for your instruction number [*insert number*] dated [*insert date*] scheduling the defects you require making good now that the rectification period has ended.

We have carried out a preliminary inspection and we are making arrangements to make good most of the items on your schedule. However, we do not consider that the following items are our responsibility for the reasons stated:

[*list, giving reasons*]

We shall of course, be happy to attend to these items if you will let us have your written agreement to pay us daywork rates for the work.

Yours faithfully

20.6 Making Good

There is no time limit for the contractor to make good the defects. The contractor must carry out its obligations within a reasonable time although that is not expressly stipulated. What is reasonable will depend on:

- The number and type of defects;
- Any special arrangements to be made with the employer with regard to access.

3.5   If the contractor fails to carry out its obligations, the architect may send the contractor a seven-day compliance notice (see Chapter 16). If the contractor still fails to make good it should be noted that the employer may have the defects made good by others and an appropriate deduction may be made from the Contract Sum. In order to achieve this the architect must get the employer's consent to issue an instruction to the contractor not to make

2.10 (2.11)   good the defects. Whether or not to consent is at the sole discretion of the

1.7   employer.

The way this works is that 14 days after the rectification period, the architect will have sent the contractor notifications of defects. There may well be a large number of them. If the employer does not want some or all of the defects making good or if the contractor will not make good, the architect must advise the employer of the right to have them made good by others. If the employer wants to take advantage of that right the architect may issue an instruction to the contractor stating which of the defects (or all of them) are not to be made good. Then the employer, or the architect if requested by the employer may get others to make good. In theory, three quotations should be obtained for the making good and the lowest accepted. In practice, that may not be feasible. The idea is that the employer must get the defects made good as cheaply as reasonably possible. The architect then has the power to make an appropriate deduction from the Contract Sum.

Appropriate deduction

An 'appropriate deduction' means a deduction which is reasonable in all the circumstances and can be calculated by reference to one or more of the following, amongst possibly other factors:

- The contract rates/priced schedule of Works/specification; or
- The cost to the contractor of remedying the defect (including the sums to be paid to third party sub-contractors engaged by the contractor): or
- The reasonable cost to the employer of engaging another contractor to remedy the defect; or
- The particular factual circumstances and/or expert evidence relating to each defect and/or the proposed remedial works.

Usually the reason for the architect, with the employer's consent, instructing the contractor not to make good will fall into one of two categories:

1. The contractor has refused or failed to make good despite being sent an instruction to do so and a seven-day compliance notice. Or, because of the exceptionally poor standard of workmanship, the employer is justified in not wanting the contractor to carry out the making good. In those cases, the appropriate deduction will be the reasonable cost to the employer of engaging another contractor to make good.
2. The employer does not want the contractor to make good because it is more convenient to the employer to have the making good carried out at some other time. In that case, the appropriate deduction will be what it would have cost the contractor to make good.

In either of the above situations, the architect must be sure to discuss the matter thoroughly with the employer before issuing any instruction. The architect must obtain a letter from the employer authorising an instruction to the contractor that making good is not required (Box 20.3).

Defects after the end of the rectification period

Where defects appear after the end of the rectification period, the contractor is, of course, still liable, because each defect is a breach of contract. The contractor must be notified but the employer is not obliged to request the contractor to make good and the employer must mitigate any loss. The

---

Box 20.3   Letter from architect to employer if some defects are not to be made good.

Dear Sir

PROJECT TITLE

I understand that you do not require the contractor to make good the following defects:

[*list*]

These defects are included in my schedule of defects issued at the end of the rectification period. In order that I may issue the appropriate instructions in accordance with clause 2.10 [*substitute '2.11' when using MWD*], I should be pleased if you would confirm the following:

- You do not require the contractor to make good any of the defects listed in this letter.
- You waive any rights you may have against any persons in regard to the items listed as defects in the above-mentioned schedule of defects and not made good.
- You agree to indemnify me against any claims made by third parties in respect of such defects.

Yours faithfully

---

contractor is not obliged to respond to a request to make good after the end of the rectification period although it may be sensible to do so. Otherwise, the employer is entitled to engage others to make good and recover all the costs from the contractor.

Although the contract does also empower the architect (as is sometimes suggested) to instruct the contractor to make good defects at the employer's expense, it is not easy to understand why the architect should ever consider *2.10 (2.11)*   doing so.

20.7   Certificate of making good

When making good has been completed, the architect must issue a certificate to that effect. The certificate has important implications with regard to the issue *2.11 (2.12)*   of the final certificate. Essentially, the final certificate cannot be issued unless the certificate of making good certificate has already been issued. The architect must not certify that making good has been achieved until satisfied (to do otherwise would be negligent).

20.8   Common problems

The architect has issued a list of defects within 14 days after the end of the rectification period. About a month later, and when the contractor has just finished dealing with the defects on the list, another serious defect comes to light. The contractor refuses to correct it on the basis that he has dealt with the defects listed and the architect must now issue a certificate of making good.

Strictly speaking, the contractor is entitled to refuse to rectify the additional defect. However, the architect and employer should offer the opportunity to do so. Because the contractor has refused, the employer is entitled to engage others. The defect, although discovered late, is a breach of contract on the part of the contractor which is responsible for its rectification. Therefore, the employer may deduct the cost of rectification from the next certificate or, if insufficient funds are available, recover the cost from the contractor through one of the dispute resolution procedures available. The contractor is responsible for all defective work for a period of six years from practical completion in the contract is 'under hand' or for 12 years if the contract is signed as a deed.

164

2.11 (2.12)

2.10 (2.11)

The contractor is entitled to the certificate of making good, because it has discharged all its obligations to make good the defects notified within 14 days of the end of the rectification period. The later defect is a latent defect because it was not visible during the rectification period and the employer's remedies are for breach of contract under the general law and not by way of the contract machinery.

The architect need not wait until the end of the rectification period to notify the contractor of defects. The defects can be notified as they arise. Can the contractor wait until the end of the period and all defects have been notified before starting the rectification process?

165

This question crops up frequently. The contract does not specify when the contractor must make good the defects. The general law says that the contractor must make the defects good within a reasonable time. What is a reasonable time will depend on all the circumstances. Matters to be taken into account will include whether the defect poses a danger or is such that if it is not dealt with urgently, it may cause other damage. Things like water leaks and failures in roof covering, electrical and gas installation defects spring to mind. Other relevant factors will be the need to allow the contractor to plan the remedial work and to make sure that work is grouped by trade. Access to the property may be another factor and sensible arrangement must be negotiated. It is quite conceivable that, if the defects are each of a trifling nature such as paint problems, scratches, cracks in wall plaster, shrinkage in timber flooring etc., it is sensible for the contractor to wait until all the defects are reported before putting all the trades concerned into the property to make good within a short timescale. Therefore, the answer to this question is yes in many cases, but urgent problems must receive urgent attention.

21 Termination

21.1 Preliminary thoughts

Employers and contractors often say that they are terminating the contract. Invariably, what they mean is that they are bringing their obligations under the contract to an end, leaving the contract in existence. Most contracts have provisions to deal with the consequences of termination. The termination clauses in MW and MWD refer to termination of the contractor's employment under the contract.

Perhaps the most important thing to say at the outset is that termination is a serious business and if the employer or the contractor contemplates termination, they should first seek legal advice. If either receives a letter from the other warning of termination, again they should take legal advice. Anyone taking a DIY approach will regret it. The consequences of a wrongful termination are severe. However, where a party honestly relies on a contract provision to terminate, although mistaken, the party probably has not repudiated the contract. Termination is best avoided if possible, and the process is fraught with pitfalls for the unwary. Quite apart from the possibility of an action for breach of contract by one party or the other if things go wrong, the employer is always placed in a difficult position as far as getting the project completed is concerned.

Even if the employer is successful in recovering costs from the contractor, the time which has been lost can never be recovered. In practice, the contractor is in an invidious position, whether it terminates its own employment or the employer does so. The formalities laid down in the termination provisions must be followed exactly if a costly dispute is to be avoided.

The architect bears a heavy burden in having to advise the employer about rights and the procedure to be followed, and the termination clauses are only too easy to misunderstand which is why legal advice should always be taken when termination is a real possibility rather than one of the options.

An important condition

6.2.1

168

169

There is an important condition attached to termination of the contractor's employment by the employer or the contractor. The termination notice must not be sent 'unreasonably or vexatiously'. In simple terms, this means that the termination notice must not be sent by either the employer or the contractor without sufficient grounds with the object of causing annoyance or embarrassment. The word 'unreasonably' in this context means 'taking advantage of the other side in circumstances in which, from a business point of view, it would be totally unfair and almost smacking of sharp practice'. The word 'unreasonably' is a general term which can include anything which can be judged objectively to be unreasonable while 'vexatious' means an ulterior motive to oppress or annoy.

21.2   If no termination in the contract

It is important for everyone to understand that, if there was nothing in the contract to allow termination of the contractor's employment, the only way to bring the employment of the contractor to an end would be at common law. It is not usually very easy to decide when all the criteria are in place to allow common law termination. in accordance with one of the ways under the general law of contract. Before looking in detail at the methods of termination under the terms of the contract it may be helpful to briefly set out the three most common ways of termination outside the contract:

Performance

This is when both parties complete their obligations under the contract and nothing remains to be done. The contract really is at an end. This of course is the best way to terminate.

Agreement

If both parties agree to bring the contract to an end, they will sign another short agreement whose sole purpose is to end the contract. This is a good way to terminate because this kind of arrangement happens when it suits both parties to end the contract and there are unlikely to be any serious disputes.

Breach of contract

A serious breach by one of the parties to a contract may entitle the other (innocent) party to treat its obligations under the contract as at an end. But it **is not** every breach of contract which entitles the innocent party to behave in this way. The breach must be 'repudiatory', i.e. it must be conduct which makes it plain that the party in breach will not perform its obligations. Until all the facts have been investigated it is sometimes difficult to say at what point a breach has occurred that is sufficiently serious to entitle the innocent party to accept the breach and bring the obligations of both parties to an end.

An option

In practice, most terminations take place under the contract terms, but the fact that there is a termination clause does not prevent either the employer or the contractor from ending obligations under the common law if they decide that it suits them better to do it that way. The employer's and the contractor's rights to terminate and the consequences are expressly stated to be without prejudice

6.3.1   to any other rights or remedies which either may possess. This means that the ordinary rights at common law are preserved and 'without prejudice' in this context means that just because there are special contract rights to terminate, the ordinary rights under common law are not affected.

Termination under the contract

In common with most standard building contracts, MW and MWD provide for either party to terminate the contractor's employment under the contract by going through a prescribed procedure. The termination provisions attempt to improve on the common law rights of the parties; indeed, the clause goes on to specify what the rights of the parties are after there has been a valid termination.

21.3   Termination by the employer

Grounds

There are five separate grounds for termination. They are that the contractor:

6.4.1.1
- Wholly or substantially suspends the carrying out of the Works before practical completion without reasonable cause; *or*

6.4.1.2
- Fails to proceed regularly and diligently with the Works before practical completion; *or*

6.4.1.3
- Fails to comply in accordance with the contract with the CDM Regulations before practical completion; *or*

6.5
- Becomes insolvent; *or*

6.6
- Commits a corrupt act.

The first three of these grounds are each described as a 'specified default'.

Procedure for a specified default

If the employer, on the architect's advice, decides to terminate the contractor's employment, the architect must ensure that the procedure is followed precisely. There is provision for preliminary notice before the right of termination is

---

> **Box 21.1   Letter from architect to contractor giving notice of default.**
>
> SPECIAL, RECORDED SIGNED-FOR DELIVERY OR DELIVERY BY HAND
>
> Dear Sir(s)
>
> PROJECT TITLE
>
> Take this as formal notice under clause 6.4.1 of the contract that you are in default in the following respects:
>
> [*insert details of the default with dates if appropriate*]
>
> If you continue the default for seven days after receipt of this notice, the employer may thereupon terminate your employment under this contract.
>
> Yours faithfully
>
> Copies: Employer
>
> Quantity surveyor [*if appointed*]

6.4.1

exercised. The architect must give notice which specifies the default and which requires it to be ended. Box 21.1 is the sort of letter which the architect might draft and it must be sent by special or recorded signed-for delivery or delivery by hand so that there is no doubt that the contractor has received it. This gives the contractor advance warning that the employer may exercise the right to terminate. In many cases, that will be sufficient to stop the default immediately. The architect must take care to obtain a receipt acknowledging delivery if delivery by hand is the chosen method.

The grounds for termination should be considered carefully, as follows:

Wholly or substantially suspends the carrying out of the Works before completion without reasonable cause

6.4.1.1

Obviously this cannot apply to a situation where the contractor has properly suspended because the employer has failed to pay the amount due to the contractor. That would be considered 'reasonable cause'. Otherwise, if the contractor completely stops work it would justify the architect issuing a default notice. It is likely that even one full day stoppage (or as the contract says 'suspension') of all work would be enough, because after receiving the notice the contractor has seven days in which to rectify the default; in this case to start working again.

4.7

Defining what 'substantially' means in this context is more difficult. The ordinary meaning is 'significantly' or 'largely'. It means to a considerable degree, not trivial. It is not enough that the contractor is not proceeding with enough

170

labour. That will fall into the next category. The contractor would have to stop working in more than half the project and probably the Works must be almost at a complete standstill.

Fails to proceed regularly and diligently with the Works

6.4.1.2

It has been remarked earlier that it is strange to find this ground for termination included when there is no express requirement for the contractor to proceed regularly and diligently. Nevertheless, the contractor's failure would entitle the employer to terminate under this ground. Whether or not the contractor is proceeding regularly and diligently is a factual question, but one which may be difficult to answer in certain circumstances; (what the law says it means is set out in Chapter 5).

171

172

At one time it used to be thought that the architect could not issue a default notice on this ground while ever there was something happening on the site, no matter how trivial. That approach was and is clearly wrong. If that is what had been meant, the contract could simply have said 'Fails to proceed'. This default is a failure to procedure in the manner and with labour and materials necessary to carry on and complete the Works by the date for completion in the contract.

Fails to comply in accordance with the contract with the CDM regulations

6.4.1.3

It is thought the failure must be clear and unambiguous to justify a termination. The contract states that each party says that it will comply with the CDM Regulations. Then it emphasises two specific things to be done:

3.9

- The employer must ensure that the principal designer carries out its duties and if the contractor is not the principal contractor, the employer must ensure that the principal contractor carries out its duties. The obligation to 'ensure' is an absolute liability to perform the duty set out. It should be noted that this is the employer's obligation, not the architect's duty.

3.9.1

173

- The contractor must comply with Regulations 8 and 15 and, if it is the principal contractor, it must comply with Regulations 12 to 14. Regulation 15 sets out the duties of contractors and regulations 12, 13 and 14 set out the construction phase plan and health and safety plan, duties of a principal contractor in relation to health and safety at the construction phase and the principal contractor's duties to consult and engage with workers, respectively.

3.9.2

Procedure for termination after default

The procedure for termination is set out in Figure 21.1. If the contractor does not cease its default within seven days of receipt of the default notice, the employer – not the architect – may serve a notice by special or recorded signed-for delivery or delivery by hand, terminating the contractor's employment.

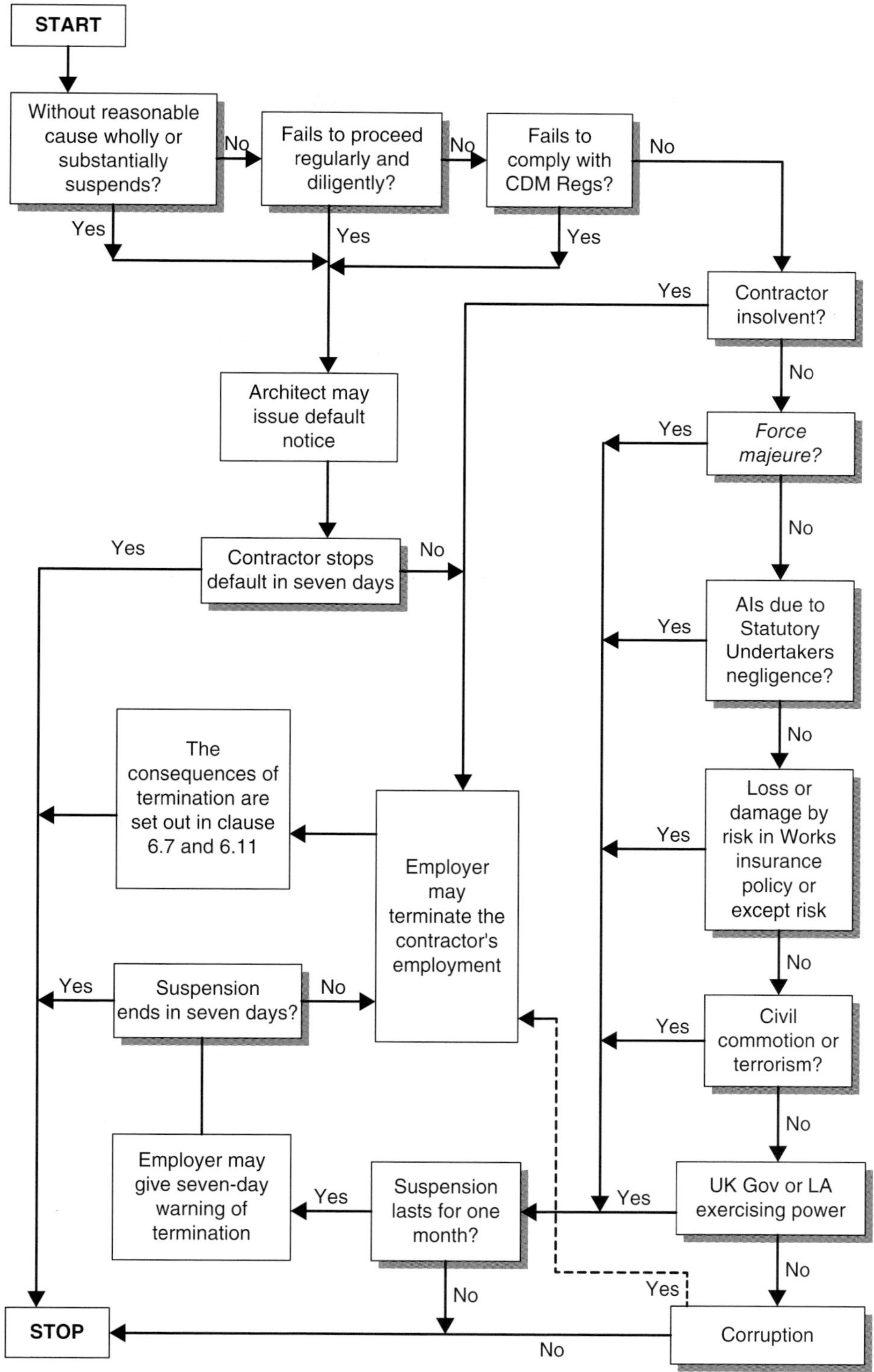

Figure 21.1 Flowchart of termination by employer.

Box 21.2   Letter from employer to contractor terminating employment.

SPECIAL OR RECORDED SIGNED-FOR DELIVERY OR DELIVERY BY HAND

Dear Sirs

I refer to the architect's letter to you dated [*insert date of architect's default letter*] and note that you have failed to end your default. [*delete the whole of this sentence if the termination is due to insolvency*]

Take this as formal notice in accordance with clause 6.4.2 [*if termination is due to insolvency substitute '6.5.1'*] of the conditions of contract that I hereby terminate your employment under the contract without prejudice to any other rights and remedies which I may possess.

You must give up possession of the site of the Works immediately and if you fail to do so, I shall instruct my solicitors to issue appropriate proceedings against you.

Yours faithfully

Copy: Architect

6.2.2	The notice operates from the date it is received by the contractor. The architect should draft a suitable letter for the employer's signature (Box 21.2).
174	

The employer has ten days from the expiry of the architect's default notice in which to serve the notice of termination on the contractor. The contract does not say what happens if the employer fails to terminate within the ten days. The procedure and timescales must be complied with very carefully otherwise the termination may be ineffective. Therefore, it seems that if the employer fails to observe the ten-day deadline, the whole procedure, including the service of a default notice by the architect, must be repeated.

Insolvency of contractor

This is the fourth ground. Termination on the ground that the contractor is insolvent is not automatic. It is by notice from the employer. The definition of insolvency has been considerably shortened from the last edition of the contract, but only because this contract refers to definitions to be found elsewhere. There is no way to put this simply. Briefly as possible, so far as the contract is concerned, insolvency is when a person or company:

- Has an administration, bankruptcy or winding-up order made or an administrative receiver, receiver or manager of property appointed or the passing of a resolution for voluntary winding up without declaration of insolvency or any other event referred to in section 113 (2) to (5) of the Housing Grants, Construction and Regeneration Act 1996; or

- Enters administration as stated in Schedule B1 of the Insolvency Act 1986; or
- Enters into an arrangement, compromise or composition in satisfaction of debts; or
- In the case of a partnership, where each partner is the subject of an arrangement or other event as noted above.

Procedure for termination on insolvency

The insolvency events are all factual matters. In some cases, e.g. where a receiver is appointed, it may be best for the employer not to terminate the contractor's employment if the receiver is willing to carry on with the company's contracts. However, if the receiver does continue the contract, care must be taken with the issue of interim certificates. In any event, specialist advice is indicated.

6.5.1

Once it is established that the contractor is insolvent as defined in the contract, the employer may give notice of termination at any time and termination takes effect on receipt by the contractor of the notice. There is no need for any prior warning notice such as is required in the case of a default. The architect should draft a suitable letter for the employer's signature (Box 21.2).

6.5.2.1

6.5.2.2

6.5.2.3

Although termination is not automatic, the contract states that as soon as the contractor becomes insolvent the consequences of termination (section 21.4 below) apply as though the termination notice had been given except for the repossession of the site by the employer and the engagement of another contractor to finish the Works. In addition, the contractor's obligation to carry out and complete the Works is suspended and the employer may take reasonable measures to secure the site and contents and the contractor must allow that to happen.

In most instances, it is sensible for the employer to terminate as soon as insolvency is established.

Corruption

The employer may terminate the contractor's employment:

- If the contractor has given or received bribes in connection with this or any other contract with the employer; or
- If the contractor commits any other offence in relation to the contract or any other contract with the employer under the Bribery Act 2010; or
- If the employer is a local authority, and the contractor commits an offence under section 117(2) of the Local Government Act 1972; or
- If regulation 73(1) of the Public Contracts Regulations 2015 apply and the circumstances set out in regulation 73(1)(b) apply. (It seems inevitable that the Public Contract Regulations 2015 will be revoked in due course).

If any of these grounds apply, the contractor's employment may be terminated by notice from the employer by special or recorded signed-for delivery or delivery by hand. The contractor must be aware that the termination may take place as a result of the corrupt actions of one of its employees or of some person acting on the contractor's behalf even if the contractor has no knowledge of the affair.

Corruption

is of course a criminal offence for which there are strict penalties, and the employer is entitled at common law to rescind the contract and/or recover any secret commissions. Legal advice is indicated.

21.4 Consequences of employer termination

If the employer successfully terminates the contractor's employment because it has wholly or substantially suspended the carrying out of the Works or failed to proceed regularly and diligently or failed to comply with the CDM Regulations or it has become insolvent or committed a corrupt act:

- The employer is entitled to engage and pay other contractors to complete the Works. The employer and any other contractor may enter and take possession of the site and of the Works. The employer is expressly given permission to use the contractor's temporary buildings, plant, tools, equipment and site materials and, therefore, the contractor may not remove them unless and until instructed to do so by the architect. The contractor's licence to occupy the site is at an end and, if the contractor fails to give up possession, it becomes a trespasser and may be evicted.

- No further sum is due to the contractor except for whatever may be due after the final accounting noted below. If a sum has already become due (for example because the architect has issued an interim certificate) it need not be paid if the employer has issued a pay less notice (see Chapter 18) or if after the date when a pay less notice could have been given, the contractor becomes insolvent. The sum need not be paid even if the contractor does not become insolvent until after termination has occurred for some other reason.

- Within three months of completion of the Works and making good of defects the architect must certify or the employer must issue a statement setting out a financial account. In this context, 'completion' refers to practical completion. but it should be noted that the contract specifically requires making good of defects to be completed in addition. What must be contained in the certificate or statement is clearly listed as all the expenses incurred by the employer in completing the Works and dealing with defects together with any direct loss and/or damage for which the contractor is liable. This loss or damage is specifically not confined to what results from the termination so it appears to include any loss or damage caused to the employer by the contractor throughout the contract period up to the date of termination. The total payments already made to the contractor must be

6.7.3.2	listed together with the total amount which would have been payable to the contractor under the terms of the contract.
6.7.3.3	■ If the total amount of employer expenses added to the payments already made is more than the amount which would have been payable to the contractor it is a debt payable from the contractor. If less than the amount which would have been payable which is less likely in practice, it is a debt payable by the employer to the contractor.
6.7.4	

21.5 Termination by the contractor

General

The contractor has a right to terminate its own employment and if it is successful in doing so, the results for the employer will be disastrous. Because of this, the architect should do everything possible to prevent it from happening. From the contractor's point of view, it will be well advised to seek legal advice as soon as it believes that it may be entitled to terminate. The practical consequences of a successful contractor termination are:

- The project will have to be completed by others, probably at a higher cost. In any event, professional and other fees will be involved, as well as other expenses.
- The completion date inevitably will be delayed.
- The employer may be faced with a liability to pay the contractor the profit it would have made if the contract had proceeded normally.

Figure 21.2 shows the procedure for termination by the contractor.

Grounds and procedure

There are six separate grounds which entitle the contractor to terminate its employment. They are that the employer:

6.8.1.1	■ Fails to pay the amount due to the contractor including VAT; *or*
6.8.1.2	■ Interferes with or obstructs any certificate; *or*
6.8.1.3	■ Fails to comply according to the contract with the CDM Regulations (these are all termed 'specified defaults'); *or*
6.9	■ Becomes insolvent; *or*
	■ Before practical completion the whole or substantially the whole of the Works is suspended for a continuous period of one month:
6.8.2.1	— Due to architect's instruction regarding variations; *or*
	— Due to impediment, prevention or default of the employer, architect or any person engaged or authorised by the employer including statutory
6.8.2.2	undertakers. These last two are termed 'specified suspension events').

It should be noted that there is no provision that these defaults or events must be 'without reasonable cause', and the only safeguard from the employer's point of view is that the contractor must not exercise its option to terminate its

6.2.1	employment 'unreasonably or vexatiously'.

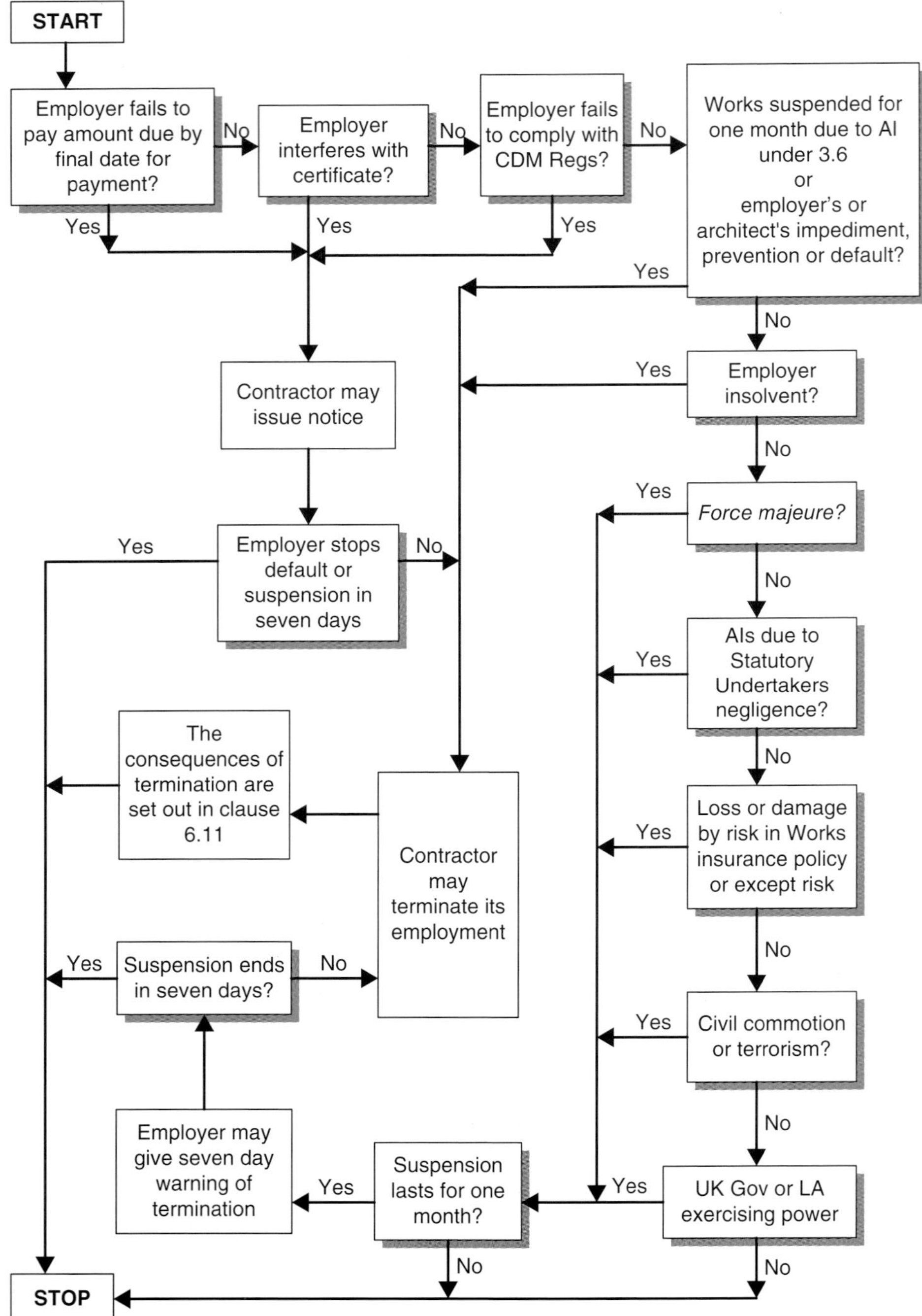

Figure 21.2   Flowchart of termination by contractor.

Procedure for specified default or suspension

The procedure for termination is set out in Figure 21.2. As in the case of employer termination, the contractor must follow the procedure precisely as a wrongful termination may amount to a repudiation of the contractor's obligations under the contract. If the contractor wishes to terminate on the basis of one of the specified defaults or suspension events, it must first send a notice to the employer (not to the architect) by special or recorded signed-for delivery or delivery by hand giving notice of its intention to terminate. That notice must specify the alleged default and require that it ends (Box 21.3). The employer has seven days from receipt of that notice in which to make good the default or stop the suspension. If the default or suspension is continued for seven days, the contractor may terminate its employment. Termination is brought about by a further notice sent by the contractor to the employer by special or recorded signed-for delivery or delivery by hand and it takes effect on receipt (Box 21.4).

6.9.1   In the case of the employer's insolvency no preliminary notice is required and the contractor may simply serve a notice of termination of its employment.

It is worthwhile considering the grounds for termination in detail because they are intended to protect the contractor against common wrongdoings by employers.

The employer fails to pay the amount due to the contractor including VAT

6.8.1.1

This ground protects the contractor's right to be paid on time, i.e. before the final date for payment and expressly includes payment of VAT. Steady cash flow is as important to the contractor as it is to the architect or to the employer and in fact this provision is quite generous to the employer. Both contractor and architect should be familiar which the fairly complex payment provisions in the contract (see Chapter 18).

The employer has to pay to the contractor, within 14 days of the date of the due date, the amounts which the architect certifies and payment is due to the contractor before that period expires. In situations where the architect has failed to issue an interim certificate on time, the contractor's payment notice will specify the amount payable and the final date for payment will be adjusted accordingly.

4.4.2

4.5.3

Any employer who receives a default notice must pay at once, and if the architect knows about it, the architect should telephone the employer and advise immediate payment and confirm this by a letter along the lines of Box 21.5.

Right from the outset, the architect must make sure that the employer understands the scheme of payments and the need to pay promptly on certificates. Where possible, financial certificates should be delivered by hand and a receipt obtained. If this is impracticable, they should be sent by special delivery and in any event by email. Many employers believe that their liability to pay does not arise until they receive an invoice from the contractor. That is incorrect.

Box 21.3   Letter from contractor to employer giving notice of default or suspension event before termination.

SPECIAL OR RECORDED SIGNED-FOR DELIVERY OR DELIVERY BY HAND

Dear Sir

PROJECT TITLE

Take this as notice under clause 6.8.1.1/.2/.3/6.8.2.1/.2 [*delete as appropriate*] of the conditions of contract of the following specified default(s)/specified suspension event(s) [*delete as appropriate*].

[*insert details of the default(s) or suspension event(s) with dates as appropriate*]

If you continue the default(s)/suspension event(s) [*delete as appropriate*] for seven days after receipt of this notice, we may on or within ten days terminate our employment under the contract.

Yours faithfully

Box 21.4   Letter from contractor to employer terminating employment after default or suspension event notice.

SPECIAL OR RECORDED SIGNED-FOR OR DELIVERY BY HAND

Dear Sir

PROJECT TITLE

We refer to the default(s)/suspension event(s) [*delete as appropriate*] notice sent to you on the [*insert date*].

Take this as notice that, in accordance with clause 6.8.3, we hereby terminate our employment under the contract without prejudice to any other rights or remedies which we may possess.

We are making arrangements to remove all our temporary buildings, plant, etc. and materials from the Works and we will write to you again within the next week regarding financial matters.

Yours faithfully

Box 21.5   Letter from architect to employer advising immediate payment.

Dear Sir

PROJECT ITLE

I refer to our telephone conversation today when I advised you to make immediate payment to the contractor of Interim Certificate No. [*insert number*] amounting to £[*insert amount*] in light of the service on you of a notice informing you that the contractor may terminate its employment under the contract.

It is safest to assume that you do not have a full seven days to pay and to pay immediately. I suggest that you either have a cheque delivered by hand to the contractor's office and a receipt obtained or payment is made electronically into the contractor's bank account. If you allow the contractor to terminate its employment, the consequences will be considerable extra cost and delay to the project.

It is essential that you always pay the amounts on Interim Certificates within 14 days of the due date which is stated on the Certificate (unless of course we have discussed, and you or I have issued, a valid pay less notice to the contractor under clause 4.5.4). It is your responsibility to make the necessary financial arrangements to ensure that funds are available to meet this obligation.

Yours faithfully

The liability arises when the interim certificate is issued and the payment period is calculated from the previous due date so that in practice it is usually several days shorter than the 14 days.

For example, if the interim valuation date is the 1 March, the due date will be seven days later on the 8 March. Therefore, the final date for payment will be 14 days later on the 22 March. If the architect takes a full five days after the due date to issue the interim certificate, i.e. on the 13 March, the employer will only have nine days, not 14, in which to pay.

Once the 14-day period for payment of a certificate has expired, the contractor may refer to adjudication in any case, although many contractors prefer to rely on their option to terminate employment.

All this of course assumes that the employer has not issued a valid pay less notice at the right time.

4.5.4

It is important to read the wording of the termination provision carefully. It does not say that the contractor may terminate if the employer does not pay the amount certified or the amount stated in a contractor's payment notice. Instead it refers to the amount due to the contractor in accordance with the relevant clause. Therefore, if the employer or the architect has issued a pay less notice, the contractor may only start the termination procedure if the employer has not paid the sum stated in the pay less notice. Whether the pay less notice is valid

4.5

7

4.7

or, if valid, whether it properly represents the amount the employer should pay can of course be dealt with in one of the dispute resolution procedures. Of course the contractor also has the right to suspend all its obligations if the employer fails to pay (see Chapter 5, Section 5.8).

6.8.1.2   ### *The employer interferes with or obstructs any certificate*

Although there is nothing in the contract which specifically states that the employer must not interfere or obstruct the issue of a certificate, there must be an implied term of the building contract to that effect. Interference or obstruction is a serious matter. In law, conduct of this kind is often referred to as 'acts of hindrance and prevention' and those kinds of acts are breaches of contract at common law.

This ground refers to any certificate, not merely financial certificates. There are other certificates that the architect is required to issue which the employer conceivably may try to prevent, such as the practical completion certificate. The architect has a duty under the contract to issue certificates. It must be made plain that the employer who tries to interfere with that duty is in breach. If, despite the warning, the employer absolutely forbids the architect to issue a certificate, the architect's duty is then to write and confirm the instructions received, setting out the consequences to the employer (Box 21.6). The architect has no duty to deliberately inform the contractor, but if the contractor

---

Box 21.6 Letter from architect to employer if employer obstructs issue of a certificate.

Dear Sir

PROJECT TITLE

I confirm that under clause 4.3 of the conditions of contract, I have an obligation to issue an Interim Certificate by [*insert date*] at the latest.

I further confirm that during a meeting with you/by email/letter/telephone call on the [*insert date*] you instructed me not to issue this Certificate. I am obliged to take your instructions, but you place me in some difficulty and I need to consider my position. Obstruction of the issue of any certificate is grounds for the contractor to terminate its employment under the contract. There will be serious financial consequences for you.

In the light of the above, I look forward to hearing that you have reconsidered your position and that I may continue to administer the contract in accordance with its terms.

Yours faithfully

suspects and terminates anyway, the architect has little option but to reveal the facts in any proceedings which may follow. Depending upon the circumstances, it may also be grounds for the architect to terminate his or her engagement by accepting the employer's conduct as repudiation at common law.

Fails to comply according to the contract with the CDM Regulations

See the comments for the equivalent ground where the employer can terminate (see section 21.3 above).

Employer's insolvency

If the employer is insolvent as understood by the contract no preliminary notice is required from the contractor, the contractor may simply give the employer a termination notice. The termination notice must be served by special or recorded signed-for delivery or delivery by hand. It takes effect on receipt by the employer. It is highly unlikely that the contractor would wish to continue and take its chance of being paid.

Once the employer becomes insolvent and even before the notice of termination takes effect, the contractor's obligation to proceed and complete the Works is suspended. This is to avoid the silly situation which would otherwise exist during this period when the contractor would be legally obliged to continue until it could terminate its employment.

Theoretically, this ground is applicable even in the case of a local authority employer, although the whole provision is drafted on the assumption that the employer is an individual or a limited company.

Causes the whole or substantially the whole of the Works to be suspended for a continuous period of one month before practical completion due to:

- *Architect's instructions regarding correction of inconsistencies or variations*
- *Impediment, prevention or default of the employer, architect or any person for whom the employer is responsible.*

If the carrying out of virtually the whole of the Works is suspended for one month for either of the reasons set out, the contractor may terminate. The first reason relates to the architect issuing instructions requiring a variation. If the contractor is delayed for one month, it will be in serious trouble. For any period up to a month, it would be entitled to loss and/or expense due to the regular progress of the Works being affected by the variation. The contract quite reasonably gives the contractor the option of termination if it foresees no quick end to the suspension and it feels unable to afford to keep the site open.

The second reason is any impediment or default of the employer, the architect, the quantity surveyor or of the employer's persons. The clause unnecessarily emphasises that the delay must not be due to the contractor's own negligence or default.

21.6   Consequences of contractor termination

6.11

If the contractor terminates its employment correctly, it is placed squarely in the driving seat. The contractor, as soon as is reasonably practicable, must prepare an account which must set out:

6.11.2.1,
6.11.2.2

- The value of work properly executed and materials properly on site for the Works and those for which the contractor is legally bound to pay. The value must be ascertained as if the contractor's employment had not been terminated, together with any other amounts due but not included.

6.11.2.3

- Any direct loss and/or damage caused by the termination (this will include loss of profit and any other costs caused to the contractor).

6.11.4

The architect is not required to issue a certificate. The contract merely requires the employer to pay to the contractor the amount properly due after taking into account everything previously paid. The payment must be made within 28 days of submission of the contractor's account.

The clause stops short of requiring the employer to pay the balance of the amount on the contractor's account; hence the use of the phrase 'amount properly due'. It is envisaged that the employer will require professional advisers to verify the contractor's account first. However, payment must be made within 28 days and it is no excuse for the employer to plead that the checking process has not been completed. It is suggested the 28 days will not begin to run until the account has been received in a form which, viewed objectively, will allow verification to take place. If the employer disagrees with the contractor's account, it is suggested that the employer should give the contractor a pay less notice no later than five days before the expiry of the 28 days. The contract does not expressly require a pay less notice in these circumstances, but it makes sense to issue one.

21.7   Termination by either employer or contractor

Either the employer or the contractor may terminate the contractor's employment if the carrying out of the whole or substantially the whole of the uncompleted Works is suspended for a minimum continuous period of one month due to:

- *Force majeure; or*
- Architect's instructions regarding variations issued as a result of negligence or default of a statutory undertaker; *or*
- Loss or damage to the Works caused by Works insurance risk or excepted risk except if the loss or damage was caused by the contractor's or contractor's persons' negligence or default; *or*
- Civil commotion or terrorist activity or threat; *or*
- Exercise by the UK Government or local authority of statutory power not triggered by the contractor's default but directly affecting the Works.

For either party to operate this clause, the Works must be lacking significant progress for the whole period as a result of the same cause. It is probable, however, that the period must be viewed as a whole.

A seven-day period of notice may be given by either party when the suspension exceeds one month. The notice must state that unless the suspension is terminated within seven days of receipt of the notice, the employment of the contractor may be terminated. If the suspension does not end, the employer or the contractor may, by further written notice, terminate the contractor's employment. Each notice must be given by special or recorded signed-for delivery or delivery by hand.

All the causes of suspension are events beyond the control of the parties. It is likely that both parties will be relieved to bring the contractor's employment to an end in such circumstances.

The consequences of termination under these grounds are the same as if the contractor had terminated due to an employer's default except that the contrac-

6.11.3   tor is not entitled to direct loss and/or damage.

21.8   Termination after loss or damage to existing structures

If termination is carried out by either party due to loss or damage to existing
5.7   structures, the consequences of termination under these grounds are the same as if the contractor had terminated due to an employer's default except that the
6.11.3   contractor is not entitled to direct loss and/or damage.

21.9   Reinstatement

It is always open to the parties to agree not to terminate even though all the necessary grounds for termination are present. Moreover, the contract expressly states that after termination has taken place the parties are free to decide to
6.3.2   reinstate the contractor's employment under the contract. That provision is unnecessary of course because the parties are always free to vary the contract in any way they choose.

21.10   Common problems

Can a termination notice be sent by the architect, and if it is the contractor terminating can its termination notice be sent to the architect?

The short answers are no and no. Architects often think that they can act for the employer about anything concerning the contract. As noted in Chapter 4, the architect does act as agent for the employer, but only in respect of those things which the contract specifies. For example, the architect can issue a pay less notice on behalf of the employer. A subsidiary question is whether the architect can issue the termination notice 'on behalf of' the employer. This is a more complicated problem. If the architect simply issues the termination without any reference to the employer, it will certainly be invalid. If the architect uses the phrase 'on behalf of' then inserts the employer's name, it is arguable that the architect is acting as agent for the employer in sending the termination notice. On the other hand, the contract sets out the powers and duties of the architect with great precision. The termination provisions say that the architect must send the default notice and the employer must send the termination notice. The contract could have said that 'the employer, or the architect on his behalf,' may issue the termination notice, but it did not do so. A contractor receiving a termination notice issued by the architect supposedly on behalf of the employer would be well advised to query the validity of the notice.

A default notice and termination notice from the contractor must be sent to the employer. Sending it to the architect is ineffective, because the contract does not say that the architect is the agent of the employer for the purpose of receiving correspondence under the contract. It is conceivable that the employer could write to the contractor authorising the architect to receive correspondence intended for the employer under the contract, but it would be a dangerous thing for the employer to do, and to comply with the contract the contractor would still have to address the document to the employer but care of the architect.

The contractor consistently produces poor workmanship and it is no better even when instructions are given to rectify it, but that is not grounds for termination.

It is unfortunate that there are contractors who are quite incapable of producing good work. By poor work is meant work which is not as specified. Obviously, when producing a tender list, the architect should take every care to check the contractors, including references from other architects. Even with the greatest care, there are cases where a contractor suffers difficulties with licence or a change of management which causes problems onsite. Sometimes, an employer will insist on using a particular contractor against the architect's advice, only to regret it later. It is true that there are no grounds for termination based simply on poor workmanship. The nearest is the contractor's failure to work regularly and diligently, but that does not cover inability to produce satisfactory work. It must be remembered that if the contractor does not carry out the Works in accordance with the drawings and specification, it is a breach or breaches of contract.

There are clauses in the contract which give the architect power to deal with poor work. The architect may instruct the contractor to rectify the defective work (see Chapter 16). The architect may also issue a compliance notice requiring the contractor to comply with the instruction within seven days. Failure to comply entitles the employer to engage another contractor to go onto the site to carry out the instruction and the architect may make an appropriate deduction from the Contract Sum. The important thing is that as soon as it becomes clear to the architect that the contractor is not promptly dealing with defects or is not dealing with them adequately, a strict regime of issuing a rectification instruction followed by a compliance notice must be followed. The architect must be relentless in carrying out inspections and issuing rectification instructions and compliance notices. If the work is really bad, the effect will be to bring most of the work to a standstill while the contractor or others is engaged in remedial work. Obviously, the contractor must not put new work on top of defective work. The situation will be reflected in zero value interim certificates. Then either the contractor will start to get things right or, if it is incapable of doing so, it might well abandon the Works, which is a ground for termination.

It is not suggested that the architect needs to take this very concentrated approach if the contractor is producing work in accordance with the specification and there are just the odd few defects here and there. But at the first sign of an inability on the part of the contractor to do good work, the architect must act firmly. One hears of projects which limp to a stage prior to practical completion before the architect really starts to treat the situation seriously. Although the fault lies primarily with the contractor who fails to produce good work, it is also the architect's fault for not picking up on the problems at an early stage.

22 Contractor's designed portion (CDP)

22.1 Principles

The contractor has no design responsibility under MW. It is sometimes thought that this can be overcome by the insertion in the specification of a paragraph giving the contractor liability for design on the basis that it does not override or modify what is in the printed form. However, giving the contractor design responsibility under MW is not easy.

179
180

Overall responsibility for design rests with the architect and the architect can only avoid this responsibility by obtaining the employer's express consent to giving the design responsibility to someone else. The Minor Works Building Contract with contractor's design 2011 (MWD) does incorporate provisions, although brief, to give the contractor design responsibility for specific items. In essence, the Contractor's Designed Portion (CDP) provisions are a very much shortened design and build contract and share some of the features of the DB contract.

181

Documents

Details of the CDP must be inserted in the second recital. A footnote advises that separate sheets may be used if the space is not sufficient to include all the items. The third recital indicates that the employer has had the Employer's Requirements prepared. When the contract is executed and this document is signed by the parties, it becomes part of the contract documents.

There is no provision, as in SBC or ICD, for the contractor to submit formal Contractor's Proposals or for a CDP price analysis.

22.2 Contractor's obligations

(2.1)

The contractor's obligations are concisely set out in the contract. Referring to the CDP, the contractor:

(2.1.1)

- Must use reasonable skill, care and diligence.
- Must complete the design for the CDP including the selection of specifications for materials, goods and workmanship to the extent that they are not

(2.1.1)

stated in the Employer's Requirements.

The JCT Minor Works Building Contracts 2016, Fifth Edition. David Chappell.
© 2018 John Wiley & Sons Ltd. Published 2018 by John Wiley & Sons Ltd.

(2.1.2)

- Must comply with the relevant parts of the CDM Regulations, particularly as they affect the designer.

- Must comply with the architect's instruction about the integration of the CDP work with the rest of the Works, subject to the contractor's consent.

(2.1.2, 3.4.2)

- Must provide the architect with two copies of drawings, details and specification reasonably necessary to explain the CDP as and when necessary to do so.

(2.1.3)

- Is not responsible for what is in the Employer's Requirements or for checking the adequacy of any design, but any inadequacy found in the Employer's Requirements must be corrected.

(2.1.4, 2.6, 3.6.1)

These provisions are discussed below.

Liability

182
(2.1.1)

For what it designed or completed, the contractor would normally have a fitness for purpose liability, but that is modified under this contract to reasonable skill, care and diligence. Therefore, like an architect, the contractor does not guarantee the result of the design, but only that reasonable skill and care was taken in its production.

(2.1.4)

183

It is expressly stated in the contract that the contractor is not responsible for what is in the Employer's Requirements, nor for verifying whether any design they contain is adequate. Without that specific provision in the contract the contractor would be responsible for checking that any design already done worked. Although the contractor has no duty to check a design included in the Employer's Requirements, if it notices an inadequacy, the Employer's Requirements must be corrected. This clearly includes any defective design. Although the clause does not state that the contractor must notify the architect on finding the inadequacy, it must be implied and, in any event, it is a matter of plain common sense. The correction of the Employer's Requirements must be undertaken by the employer or by the architect on the employer's behalf. Any delay in correcting the problem would entitle the contractor to an extension of time if it delayed the completion date. Problems may occur if an inadequacy exists in the Employer's Requirements, but the contractor completes the design without noticing or checking. Although the contractor specifically has no liability to check, the architect may try to contend that it must have been obvious to the contractor when completing the design. It would be for the architect to demonstrate that the contractor must have noticed the inadequacy.

(2.2.1)

If materials, goods or standards of workmanship in the CDP are not described in the contract documents, they must be of a standard which is appropriate to the Works. Effectively, that means that if there is no description in the Employer's Requirements, the contractor has the responsibility of producing something appropriate.

Integration of the CDP

(2.1.2)

Integration with the rest of the Works is a common problem area. The architect is given power to issue directions for integration of the CDP design. The contractor will often contend that the architect's instructions for integration unavoidably result in additional work and will, therefore, seek a variation. The principles are straightforward although application to particular circumstances may need care. There are four possible situations:

- It is a matter for the contractor to allow in its proposals for the proper integration of the CDP with the rest of the design if the invitation to tender is supported by clear documents.
- The contractor probably has a claim for any additional costs resulting from the architect's directions on integration if the invitation to tender is not supported by sufficient information to enable the contractor to properly design the interface between the CDP and other work.
- The architect must issue directions on integration and the contractor has a claim for additional costs if the architect subsequently issues instructions that affect the CDP.
- The contractor must bear its own costs if it is obliged to alter the CDP in order to correct its own error. In such an instance the architect will probably have to issue some directions about the integration of the corrected CDP.

(3.4.2)

The architect may not issue any instruction which affects the design of the CDP work unless the contractor consents. It will be a matter to consider in each case whether the architect's directions on integration affect the design of the work.

Contractor's information

(2.1.3)

(2.1)

The contractor is obliged to provide the architect with copies of documents, details and specifications reasonably necessary to explain the CDP. The architect is entitled to request any related calculations or other information. The architect may specify the format for drawings and procedures for submission of design information in the Employer's Requirements. As a minimum, the contractor must submit the documents so as to allow the architect seven full days from the date of receipt to consider them and make any comments before the contractor uses them for construction.

22.3   Inconsistences and divergences

(2.5.2)

(2.2.1)

If there are inconsistencies within or between the individual CDP documents prepared by the contractor, the contractor must correct the inconsistency at its own cost. However, that is made subject to the architect's approval of the contractor's proposed correction, which means that it must be to the architect's reasonable satisfaction. The contractor ought to request the approval in writing or, at least, confirm it back to the architect before proceeding.

(2.6.1)

(2.6.2)

(2.1)

184

If the contractor becomes aware that there is a divergence between statutory requirements and any other contract document or architect's instruction, it must immediately notify the architect. If the contractor does that, it will not be liable for any divergence between the Works and statutory requirements. However, that assumes that the contractor has carried out the Works in accordance with the contract documents and that no CDP work is involved Obviously, if the reason for the divergence is simply that the contractor has carried out the construction wrongly, the contractor will be responsible for correcting it at the contractor's own cost. It is also obvious that if the divergence concerns the CDP work, the contractor must correct it. Unlike the position with inconsistencies, the contract does not say that the architect must approve the contractor's proposal to correct the divergence, but it is clear that it cannot be carried out unless the architect does approve it. The contract quite specifically says 'If the contractor becomes aware…'; the contractor is not obliged to check through the documents to see if there are divergences. This can lead to difficulties if the contractor does not become aware of a divergence and it is in the contract documents. A divergence in the CDP work is the contractor's problem.

22.4 Variations

(3.6.1)

(3.4.2)

3.6

The contract makes clear that an instruction requiring a variation to the CDP can only be issued in respect of the Employer's Requirements. The employer cannot issue an instruction directly about the CDP design. Therefore, it is for the architect to instruct a change to the Employer's Requirements to which the contractor responds by altering its design.

As noted above, the architect may not issue an instruction which affects the design of the CDP unless the contractor consents. In order to comply, the architect would first have to issue an instruction, which may affect the design of the CDP, in draft to the contractor. That is because the wording states that the architect must not *issue* that kind of instruction without consent. Therefore, it is not a case of the architect issuing an instruction and the contractor objecting. The contractor is placed firmly in the driving seat.

Valuation of variations in CDP work is to be carried out as any other variation (see Chapter 17).

22.5 Other matters

MWD does not require the contractor to take out professional indemnity insurance. Often, the design will be done by others sub-contracted to the contractor and they will have ongoing professional indemnity insurance. The employer may need the advice of an insurance broker whether a clause should be written into the contract. If insurance is required, the contractor must maintain it for the appropriate limitation period under the contract: six years if the contract is executed under hand, twelve years if it is a deed.

The contractor owns the copyright in its designs. There is no express provision in the contract for the employer to receive a licence to use the contractor's design, but a term to that effect would be implied.

There are no provisions to limit the contractor's liability nor is there any requirement for as-built drawings in respect of the CDP. It may be thought prudent to require as-built drawings for future reference. If so, something must be written in to the Employer's Requirements.

22.6 Common problems

Is the contractor just paid for construction or for the design as well?

It is certain that the contractor will have priced for the designing of the CDP as well as building it. That is particularly the case if the contractor engages an architect or other design professional to carry out the design work. Even if the contractor because of its staffing feels able to carry out the design itself, it will include something for the design. It is a mistake for the contractor to simply include the design price as part of its general pricing for CDP work. It is essential that the price for design is shown separately; perhaps at £x per hour. This will enable the contractor to claim payment even if the architect varies the Employer's Requirements but then cancels the instruction after the contractor has had the design revised but before it has been constructed. Otherwise, the architect may refuse to value the aborted instruction because the contractor has not actually carried out the variation.

The architect has sent an instruction in the form of a drawing showing an amendment to the design of the CDP Works.

The contractor need not comply with that instruction because it requires the contractor's consent before issue but also because the architect has no power to issue an instruction directly altering the CDP Works. An instruction must be given to vary the Employer's Requirements. Any consequent change in the design is a matter for the contractor.

(3.4.2)

(3.6.1)

In many instances, of course, the contractor will simply comply with the architect's drawing. In that case an interesting question arises as to whether it is the contractor who is responsible for the efficacy of the altered design or whether it is the architect's responsibility because there is no doubt that the architect did design the alteration. There is no simple answer but it is likely that the architect will be liable for any fault in the design to the extent that it causes some loss or expense to the employer.

23 Dispute resolution procedures

23.1 General

The architect will be expected to advise the employer about what to do if there is a dispute which cannot be resolved by negotiation. Many disputes arise because one or other of the parties misunderstands something in the contract. Hopefully readers of this book will have a good understanding of the contract, but if, nevertheless a dispute does arise which cannot be easily resolved, the employer will be looking for some good general advice on what to do next. Actually, neither the contractor, nor the employer, nor any of the construction professionals can be expected to know very much about the detail of dispute resolution procedures. If any of the procedures are employed, the employer or the contractor should engage specialists in the particular procedures. It is most unwise for construction professionals to attempt to give anything other than basic advice in these areas. What the employer will be looking for from the architect is the choices of dispute resolution and which one might be most appropriate. What follows goes beyond that on the basis that the architect should always know rather more than absolutely necessary.

23.2 Choice

The contract provides for four systems of dispute resolution.

Mediation

The first system is mediation. It cannot be used unless the parties agree. It is briefly considered at the end of this chapter. The other systems are adjudication, arbitration and legal proceedings. People understandably get very confused about when each system may be used. The way it works is this:

Adjudication

Every construction contract which does not involve a residential occupier must have an adjudication clause. This contract is commonly used where the

---

Article 6,
clause 7.2

employer is a residential occupier. If that is the case, the adjudication provisions in the contract can be deleted. If, after hearing the pros and cons, the employer wishes to retain the adjudication clause, that is fine, but a residential occupier does have the choice. Obviously, if the contract is signed and the adjudication clause has been deleted, the employer cannot subsequently decide to put it back in the contract and vice versa. What is a residential occupier? It is someone living in their own property who wants work doing to that property. It could be someone living in a flat who wants to have a house built to live in. However, an employer living in a house who wants another house constructing in part of the garden so that it can be sold off for the benefit of the employer is not a residential occupier. If the employer is not a residential occupier the adjudication clause must be retained. If the clause is retained, either employer or contractor can refer a dispute to adjudication at any time. Adjudication is intended to be a quick process. Some people call it rough justice. The decision is binding, but only until the dispute is dealt with by one of the other two systems. If the dispute is never referred to one of the other systems, the adjudication decision is effectively permanently binding. The parties can of course agree if they wish that the adjudicator's decision will be finally binding.

Arbitration or legal proceedings: Choice of procedures

The parties must choose which of the other two systems (arbitration or legal proceedings) they wish to have as their main dispute resolution system. They can refer a dispute straight to this system without going through adjudication if they wish.

Therefore, a party with a dispute may refer it for a decision by adjudication and, if dissatisfied (or if the other party is dissatisfied) may then refer the same dispute to either arbitration or legal proceedings (whichever is in the contract). Alternatively, the party with a dispute may ignore adjudication and refer the dispute immediately to either arbitration or legal proceedings (whichever is in the contract).

There are advantages and disadvantages with all these systems and the rest of this chapter attempts to highlight the pros and cons and give a general indication of the processes involved in each system. However, it cannot be emphasised too strongly that for each of these systems, the parties need to be properly represented by specialists in the particular field. It may be tempting to think that all the employer or contractor needs to do in each case is to telephone their solicitor. It must be understood that solicitors, like other professions, specialise. Many solicitors carry out litigation (legal proceedings), but there are all kinds of litigation and a solicitor is required who has a good knowledge of construction law. So far as adjudication and arbitration are concerned, some solicitors seem to be uncomfortable with these processes and, unless a solicitor is available who specialises in construction, a party may get better representation from an architect or quantity surveyor who has specialist knowledge and experience in the field. There are a good many independent consultants practising in these areas.

23.3   The Construction Act 1996

In 1996 the Housing Grants, Construction and Regeneration Act (commonly called the Construction Act) (the Act) was enacted (in Northern Ireland the Construction Contracts (Northern Ireland) Order 1997 is virtually identical to Part II of the Act). The Act was amended by the Local Democracy, Economic Development and Construction Act 2009 (the Act to the same effect in Northern Ireland is the Construction Contracts (Amendment) Act (Northern Ireland) 2011). S.108 of the Act expressly introduces a contractual system of adjudication to construction contracts.

Article 6,
clause 7.2

Every construction contract must include provision for adjudication which complies with the Act unless the contract relates to work on dwellings occupied or intended to be occupied by one of the parties to the contract. MW and MWD, in common which other standard forms, incorporates the requirements of the Act. Therefore, all construction Works carried out under these forms are subject to adjudication even if they comprise work to a dwelling house unless, as noted earlier, the adjudication provisions are deleted.

Every construction professional and every contractor should have a basic knowledge of the Act. In essence s108 of the Act provides that:

- A party to a construction contract has the right to refer a dispute under the contract to adjudication at any time.
- An adjudicator should be appointed and the dispute referred within seven days of the notice of intention.
- The adjudicator must make a decision in 28 days or whatever period the parties agree.
- The period for decision can be extended by 14 days if the referring party agrees.
- The adjudicator must act impartially.
- The adjudicator may use his or her initiative in finding facts or law.
- The adjudicator's decision is binding until the dispute is settled by legal proceedings, arbitration or agreement.
- The adjudicator is not liable for anything done or omitted in carrying out the functions unless in bad faith.

If the contract does not comply with the Act, the Scheme for Construction Contracts (England and Wales) Regulations 1998 as amended (the Scheme) will apply (In Northern Ireland it is the Scheme for Construction Contracts in Northern Ireland Regulations (Northern Ireland) 1999 as amended). The Scheme contains detailed provisions for the procedure. Under MW and MWD

7.2

the procedure is stipulated to be the Scheme.

23.4   Adjudication in general

At any time

The right to refer to adjudication 'at any time' means that adjudication can be commenced even if legal proceedings (and arbitration) are in progress about the

185   same dispute. Adjudication can be sought even if repudiation of the contract has taken place. A dispute may be referred to adjudication and the adjudicator may give a decision even after the expiry of the contractual limitation period. Of

186   course the referring party runs the risk that the respondent will use the limitation period defence. In which case the claim will normally fail.

Adjudication decision

In most cases, it seems that the parties accept the adjudicator's decision and do not take the matter further. Even where there are challenges through the courts against the enforcement of an adjudicator's decision, the challenge is concerned with matters such as the adjudicator's jurisdiction or whether the adjudicator complied with the requirements of natural justice, not whether the adjudicator's decision was correct. The courts cannot interfere with the adjudicator's decision, no matter how obviously wrong, provided that the adjudicator has the jurisdiction (i.e. properly appointed and carrying out the duties correctly) to

187   answer the questions posed by the referring party.

The parties must comply with the adjudicator's decision following which, if they are not satisfied, either party may instigate proceedings through arbitration or legal proceedings (whichever is chosen). It is important to remember that, in doing so, the parties are not appealing against the decision of the adjudicator and the arbitrator or court will ignore the adjudicator's previous deci-

188   sion in arriving at an award or judgment respectively. However, the same dispute cannot be referred to adjudication twice. Once decided by adjudication, a disgruntled party must go to arbitration or legal proceedings (whichever is in the contract). The whole idea behind adjudication is that it is a quick, cheap method of resolving disputes. That is why, unless the parties agree, there is no provision for the winner to recover legal costs from the other side. Parties are encouraged to seek modest representation. Indeed, the original idea was probably that employer and contractor could simply argue their own cases. However, the rate at which the courts have been called upon to make decisions about various aspects of adjudication means that a party is very unwise if they do not seek representation for all but perhaps the very simplest and financially modest disputes.

Article 7,   If the parties wish to have a final and binding decision rather than submit a
clause 7.3,   dispute to adjudication, they have a choice between arbitration and legal
Schedule 1   proceedings. It should be noted that legal proceedings will apply unless the
Article 8   contractor particulars are completed to show that arbitration is required.

23.5   Pros and cons

The advantages of arbitration

- *Speed*: A good arbitrator should dispose of most cases in months, not years.
- *Privacy*: Only the parties and the arbitrator know the details of the dispute and the award.

- *The parties decide*: The parties can decide the identity of the arbitrator, the timescales, the procedure and the location of any hearing.
- *Expense*: Theoretically, it should be more expensive than litigation, because the parties (usually the losing party) have to pay for the arbitrator and the hire of a room, but in practice the speed and technical expertise of the arbitrator usually reduces costs.
- *Technical expertise of the arbitrator*: The fact that the arbitrator understands construction should shorten the time schedule and possibly avoid the need for expert witnesses if the parties agree.
- *Appeal*: The award is final, because the courts are loath to consider any appeal.

Disadvantages of arbitration

- In theory, it is more expensive because the parties (usually the losing party) pay the cost of the arbitrator and the hire of a room for the hearing.
- If the arbitrator is not very good, the process may be slow and expensive.
- The arbitrator may not be an expert on the law, which may be a major part of the dispute.
- Parties who are in dispute may find it difficult to agree about anything. Therefore, the arbitrator may be appointed by the appointing body and the procedure, the timing and the location of the hearing room may be decided by the arbitrator with the result that neither party is satisfied.

Advantages of legal proceedings

- The judge should be an expert on the law.
- The Civil Procedure Rules require judges to manage their caseloads and encourage pre-action settlement through use of the Pre-Action Protocol.
- Cases can reach trial quickly.
- The claimant can join several defendants into the proceedings to allow interlocking matters to be decided.
- Costs of judge and courtroom are minimal.
- A dissatisfied party can appeal to a higher court.

Disadvantages of legal proceedings

- Even specialist judges know relatively little about the details of construction and often have to rely on expert witnesses who can be costly.
- Parties cannot choose the judge, who may not be very experienced in construction cases.
- Costs will be added because expert witnesses or a court-appointed expert witness may be needed to assist the judge.
- Although matters have significantly improved in recent years, cases can sometimes take a long time to resolve. Lengthy timescale and complex processes may result in high costs.
- Appeals may result in an unacceptable level of costs.

Arbitration

Arbitration is probably still the most satisfactory procedure for the resolution of construction disputes and employers would be advised to complete the contract particulars accordingly. Where the parties have agreed that the method of binding dispute resolution will be arbitration, a partly who attempts to use legal proceedings instead will fail in a costly way if the other party asks the court to grant a stay (postponement) of legal proceedings until the arbitration is concluded. The court has no discretion about the matter and the successful party will claim its costs. The result is not only that the party intent on legal proceedings will have to revert to arbitration, but it will have to pay the other party's legal costs incurred in opposing the legal proceedings.

The remainder of this chapter sets out the procedures in more detail for the benefit of the reader who wishes to know a little more about dispute resolution. However, it must be borne in mind that reading what follows is not sufficient, in itself, to qualify anyone to attempt to represent a party in adjudication or arbitration.

23.6   Adjudication in detail

The contract provisions

Although in most instances it will be the contractor that initiates adjudication, there is nothing to stop the employer from doing so. For example, the employer may seek adjudication to ask the adjudicator if the architect has over-certified or made an excessive extension of time.

The architect is not a party to the contract and, therefore, cannot be the respondent to an adjudication under MW or MWD (of course the architect can be a party to an adjudication with the employer under the architect's appointment document). Architects can obviously act as witnesses, but they have no duty to run an adjudication on behalf of the employer. Representing someone in an adjudication usually calls for some degree of skill and experience which most construction professionals, acting in the normal course of their professions, will not readily acquire. Where the dispute is other than very straightforward or where one party has retained the services of a legal representative, the other party is well advised to do likewise.

It used to be thought that only disputes arising 'under' the contract may be referred. The situation has now changed and no distinction should be made between references to disputes 'under,' 'arising out of' or 'in connection with' the contract.

Adjudication is rapidly replacing arbitration as the standard dispute resolution process even though it is rather rough and ready. The reason is probably because it is a very quick process. Use of the Scheme is made subject to the adjudicator and nominating body being those stated in the contract particulars.

The Scheme: Notice of adjudication

To start an adjudication, any party to a construction contract may give to all the other parties (there is usually only one other party) a written notice of an intention to refer a dispute to adjudication. The notice must describe the dispute and the parties involved. It must give details of the time and location, the redress sought and the names and addresses of the parties to the contract.

The notice is the trigger for the adjudication process and it is also one of the most important documents. Great care must be taken in its preparation because the dispute which the adjudicator is entitled to consider is the dispute identi-191 fied in the notice of adjudication. The dispute cannot be changed later by the referring party, although it can be elaborated and more detail provided and possibly widened especially by the nature of the defence. Unless the parties agree otherwise, only one dispute may be referred although in some circum-192 stances it may involve several contracts. The adjudicator has the power to decide that several contracts are simply variations of one contract or to deal with disputes arising under a framework agreement together with all its works 193 packages.

If the referring party states the dispute as being the amount due in an architect's certificate and omits to request the adjudicator to order payment of that amount, the adjudicator is entitled to order payment of the amount 194 in the decision. If an adjudicator is asked to decide a claim for sums due on an interim valuation and, after making the valuation, decides the sum is due from the claimant to the other party, he can order that payment. However, 195 if an adjudicator is asked to decide if £10,000 is due to the referring party, the adjudicator will only be able to decide whether the £10,000 is due or not. If the adjudicator thinks that only £9,500 is due, the decision will have to state simply that £10,000 is not due. In order for the adjudicator to decide something different from the amount claimed, the referring party must include some such words as '£10,000 or such other sum as the adjudicator 196 decides'.

However, the adjudicator may take into account any other matters which both parties agree should be within the adjudication's scope. Moreover, the adjudicator is expressly empowered to take into account matters which the adjudicator considers are necessarily connected with the dispute. To take a simple example, it is probably essential for an adjudicator to decide the extent of extension of time allowable, even if not asked, before deciding about the amount of liquidated damages properly recoverable. The express empower-197 ment merely puts into words what would be the legal position in any event. The responding party is entitled to put forward any defence even if not previously 198 relied upon.

The Scheme: Appointment of the adjudicator

The procedure for selecting an adjudicator is, unfortunately, fairly complex (see flowchart Figure 23.1).

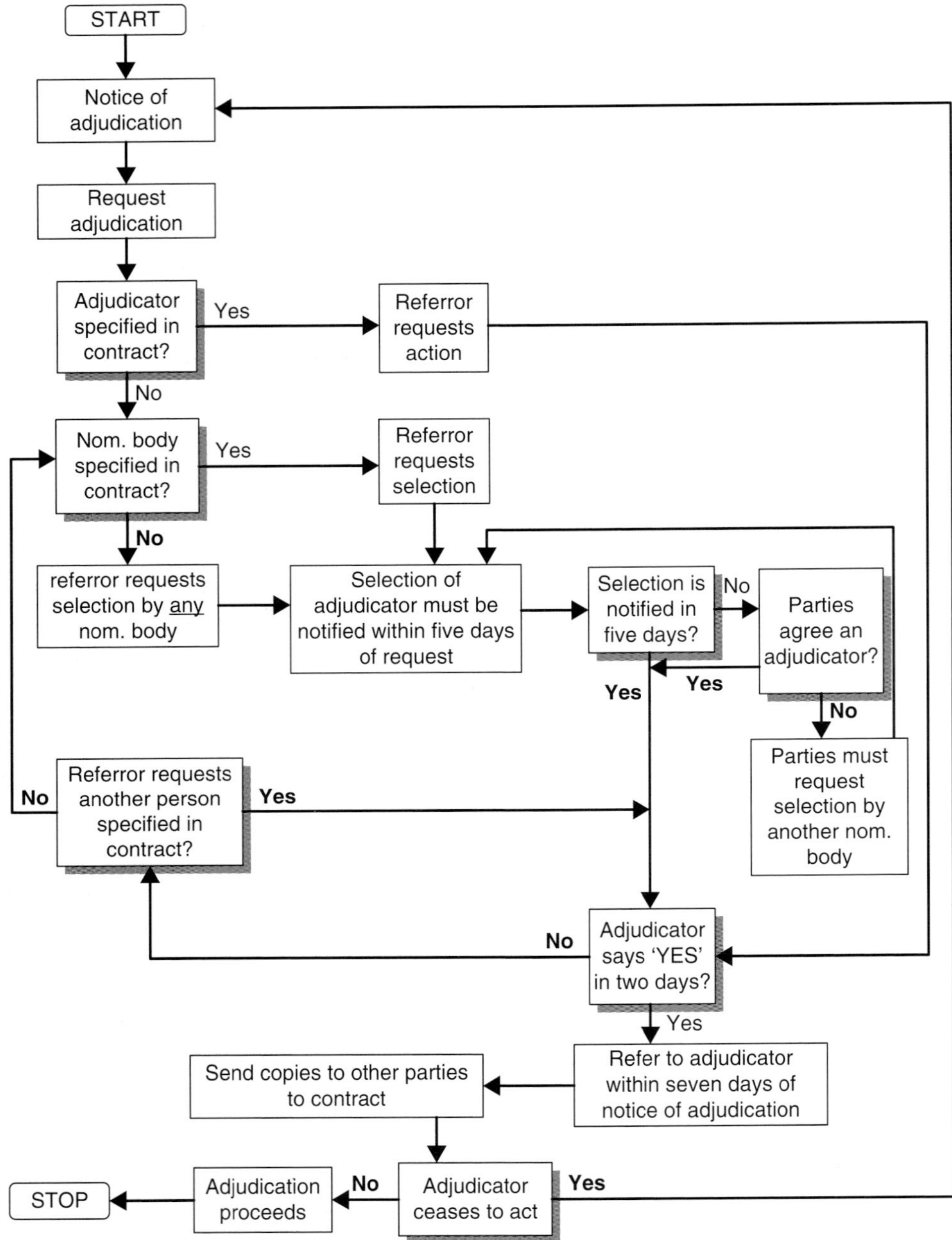

Figure 23.1   Selection of an adjudicator.

Parties may agree

The main point is that the parties may agree the name of an adjudicator after the notice of adjudication has been served. If the parties do agree, they have the best chance of getting an adjudicator who has the confidence of both parties. The chances of agreement tend to be rather small.

Named adjudicator

If there is a person named in the contract, that is the person who must first be asked to act as adjudicator. There are problems with having a named person:

- The person may be away or ill or even dead when called upon to act;
- The person's expertise may be unsuitable for the particular dispute;
- Pressure of work may force that person to decline.

If an adjudicator is named in the contract, but cannot act or does not respond, the referring party has three options:

- Ask any other person specified in the contract to act;
- Ask the adjudicator nominating body in the contract to nominate;
- Ask any other nominating body to nominate.

This procedure is simply a clarification of existing options. The nominated adjudicator has two days from receiving the request in which to accept. The adjudicator must be a single person and not a company. Therefore, a firm of quantity surveyors cannot be nominated although one of the directors or partners can be nominated. The nominating body has five days from receipt of a request to communicate the nomination to the referring party. Invariably, a nominating body will also notify the respondent and it is surprising that the Scheme does not expressly require that notification.

Nominating body

If there is no named person or if that person will not or cannot act, the referring party must ask the nominating body stated in the contract particulars to nominate an adjudicator. There are problems with this approach also. Not all adjudicators are of equal capability. Indeed, some of them are not good at all. Some have a tenuous grasp of the law while some others believe that the adjudicator's job is to make decisions according to their own notions of fairness without reference to law. If the referring party asks for a nomination, both parties are stuck with the result unless they agree to revoke the appointment. However, for the reason already stated, such an agreement is unlikely.

The nominating body is to be stated in the contract particulars. Five bodies are listed and four of the bodies should be deleted. If there is no adjudicator named and no body is selected, the referring party may choose any one of the bodies to make the appointment. If there is no list of nominating bodies, perhaps because all of them have been inadvertently deleted, the referring party is free to choose any nominating body, including bodies not on the list, to make the appointment.

A nominating body is fairly broadly defined in the Scheme as a body which holds itself out publicly as a body which will select an adjudicator on request. The body may not be what is referred to as a 'natural person', i.e. a human being, nor one of the parties.

In the event that the nominating body fails to nominate within five days, the parties may either agree on the name of an adjudicator or the referring party may request another nominating body to nominate. In either case the adjudicator has two days to respond as before.

If the nominating body took five days to nominate and the adjudicator took two days to respond, only to decline, there would be no time to re-nominate and the process would have to re-commence from the notice of adjudication. In practice, the referring party will apply to the nominating body in the contract particulars who will first find an adjudicator who is willing to act before notifying both parties of the name of the adjudicator within about three or four days of the application.

Objections

If either party objects to the adjudicator, it will not invalidate the appointment nor any decision reached by the adjudicator. Many people think that if one of the parties registers an objection to the adjudicator, the process will come to an end. That is completely wrong. Each party can decide the extent to which it wishes to participate in the adjudication. If a party objects to something, it should make quite clear that its participation in the adjudication process is without prejudice to that objection and to the party's right to refer the objection to the courts in due course. If an objecting party continues the adjudication without further comment, it may well have lost its chance to object later.

Adjudicator ceasing to act

The Scheme has detailed provisions for the situation where the adjudicator resigns or the parties revoke the appointment. The adjudicator may resign at any time on giving notice in writing to the parties. No notice period is specified and so the adjudicator can simply write to the parties saying that he has resigned. The referring party may serve a new notice of adjudication and seek the appointment of a new adjudicator as noted earlier. If the new adjudicator requests and if reasonably practicable, the parties must make all the documents available which have been previously submitted. That is an odd provision in the Scheme, because it is difficult to see how a replacement adjudicator could proceed without that information.

An adjudicator, who finds that the dispute is essentially the same as a dispute which has already been the subject of an adjudication decision, must resign. If so, or if there is significant variation of a dispute from what was referred in the referral notice, so that the adjudicator is not competent to decide it, the adjudicator can specify a reasonable fee. Parties will find it difficult to challenge the amount of fees determined by an adjudicator unless the adjudicator can be shown to have acted in bad faith.

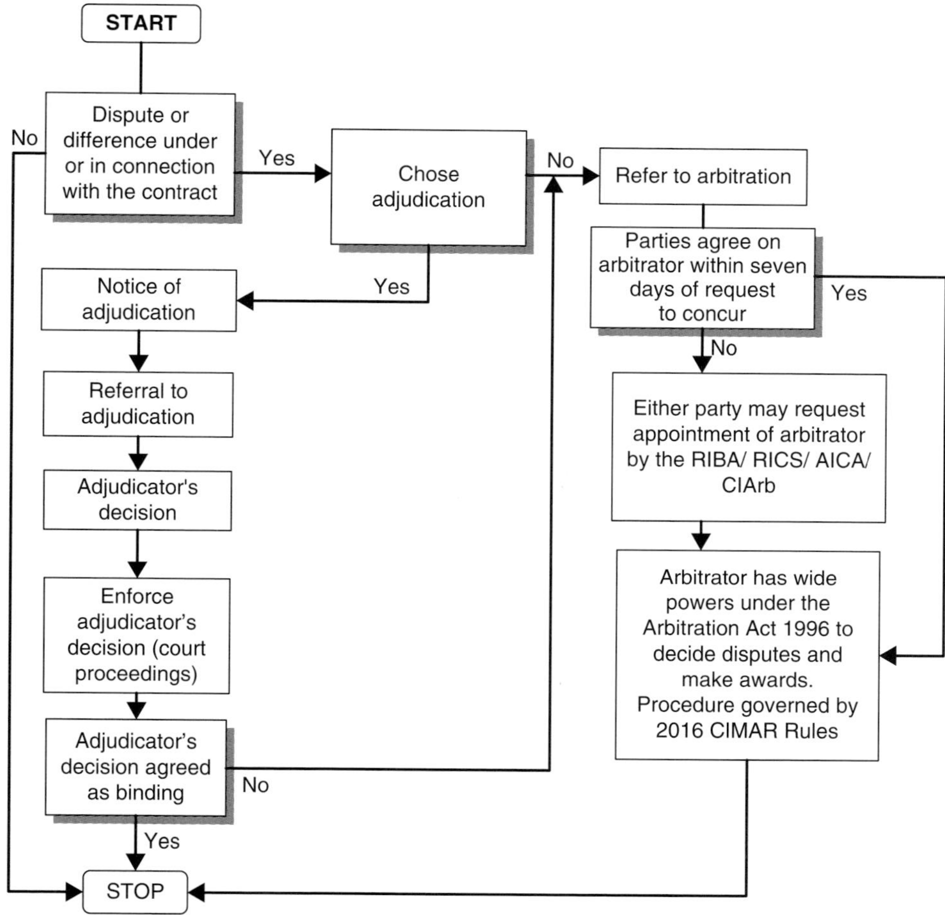

Figure 23.2   Procedure under the Scheme and in relation to arbitration.

The Scheme: Procedure

The procedure is indicated by a flowchart in Figure 23.2. A request for the appointment of an adjudicator must include a copy of the notice of adjudication. This is to assist those making the nomination and the prospective adjudicator so that a suitable person is nominated. Time is short.

Referral, response and decision

The referral notice, which is the referring party's claim, must be accompanied by relevant parts of the contract and whatever other evidence the referring party relies upon in support of the claim.

The referring party must submit the dispute in writing to the adjudicator, with copies to each party to the dispute, no later than seven days after the notice of adjudication. This submission is known as the 'referral notice'. Looking at the procedure for appointment of the adjudicator, it is clear that the timetable is tight. It appears that provided the referral notice is delivered to the adjudicator and the other party by day seven, it may be acceptable in some cases if some of the accompanying documents are not delivered until day eight.

The Scheme does not say that the respondent may reply to the referral notice, but, in practice, the adjudicator is obliged to allow a reasonable period for the reply. The length of time is a matter for the adjudicator to decide, but in view of the restricted overall period for the decision, it is likely that somewhere between seven and 14 days is the most which any respondent can expect. The adjudicator must reach a decision twenty-eight days after the date of receipt by the adjudicator of the referral notice. The period may be extended by 14 days if the referring party consents or, if both parties agree, for any longer period. The adjudicator must deliver a copy of the decision to the parties as soon as possible after the decision has been reached.

If the adjudicator does not comply with this timetable in reaching the decision, either party may serve a new notice of adjudication and request a new adjudicator to act.

Enforcement of the decision

If one of the parties fails to comply with the adjudicator's decision, the other may ask the court to enforce the decision. A court will normally enforce the decision. In enforcement proceedings, the court is not being asked to decide whether the adjudicator got the decision right, but only whether he had the power to make the decision. If the adjudicator was properly appointed and he conducted the adjudication properly, the court will enforce the decision.

The Scheme: Adjudicator's powers and duties

Several disputes

Unless the parties agree, the adjudicator may only adjudicate at the same time on one dispute between the parties. If the parties agree, the adjudicator may adjudicate on more than one dispute under the same contract. The adjudicator may deal with related disputes on several contracts even if not all the parties are parties to all the disputes, provided they all consent. Moreover, the parties may agree to extend the period for decision on all or some of the disputes. It is clear that multiple dispute procedures bring their own complications for which the Scheme, wisely, does not try to legislate. For example, it is not clear whether multiple disputes, certainly under different contracts, must be adjudicated on during one big adjudication. Where there are different contracts and the parties vary from one contract to another, it will be a matter of discussion and agreement whether the adjudicator should conduct separate adjudications at the same time.

Acting according to law

The adjudicator's duties are to act impartially in accordance with the relevant contract terms, to reach a decision 'in accordance with the applicable law in relation to the contract' and to avoid unnecessary expense. Some adjudicators seem to be unaware of their obligations to apply the law to their decisions and decisions are made on the basis of the adjudicators' idea of fairness, moral rights or justice. Fortunately, there are also some very good adjudicators with a clear understanding of their roles.

Summary of powers

The adjudicator is given some very broad and some very precise powers:

- To take the initiative in ascertaining the facts and the law;
- To decide the procedure in the adjudication;
- To request any party to supply documents and statements;
- To decide the language of the adjudication and order translations;
- To meet and question the parties;
- To make site visits, subject to any third party consents;
- To carry out any tests, subject to any third party consents;
- To obtain any representations or submissions;
- To appoint experts or legal advisers, subject to giving prior notice;
- To decide the timetable, deadlines and limits to length of documents or oral statements;
- To issue directions about the conduct of the adjudication.

Failure to comply

If either party does not comply with the adjudicator's directions, the adjudicator has power:

- To continue the adjudication despite the failure.
- To draw whatever inferences the adjudicator believes are justified in the circumstances.
- To make a decision on the basis of the information provided and to attach whatever weight to evidence submitted late the adjudicator thinks fit.

A party may have assistance or representation as deemed appropriate, but in the case of oral evidence or representations, for example, in the case of a meeting or site visit, only one person may represent each party unless the adjudicator decides otherwise. In practice, it is seldom that the adjudicator will enforce a limit on representation and I have seen meetings where each party is represented by a barrister and two solicitors as well as an expert witness.

Final certificate

Sometimes, a contract may provide that a decision or certificate is final and conclusive in certain circumstances. Unless that is the case, the adjudicator has power to open up, revise and review any decision or certificate given by a person named in the contract. It is worth noting that to be exempt from revision by the adjudicator the decision or certificate must be stated to be both final and conclusive. A contract which simply state that a certificate is conclusive is open to review. On that basis, the final certificates under SBC, IC and ICD are exempt because they are called 'final' and stated to be conclusive. It should be noted that the final certificates under MW and MWD are neither expressed to be final nor conclusive and they can be reviewed by the adjudicator.

Power to order payment

The adjudicator is also given power to order any party to the dispute to make a payment, its due date and the final date for payment and to decide the rates of interest, the periods for which it must be paid and whether it must be simple or compound interest. In deciding what, if any interest must be paid, the adjudicator must have regard to any relevant contractual term. To 'have regard' to a contractual term is a rather loose phrase which probably means little more than to give attention to it. It falls short of the need to actually comply with it.

Information submitted

The adjudicator must consider relevant information submitted by the parties and if the adjudicator believes that other information or case law should be taken into account, it must be provided to the parties and they must have the opportunity to comment. Neither the adjudicator nor any party to an adjudication may disclose information, noted by the supplier as confidential, to third parties unless the disclosure is necessary for the adjudication.

The Scheme: The adjudicator's decision

The adjudicator must reach the decision within 28 days (or within any validly extended period) after receipt of the referral. The decision must be delivered to each party as soon as possible after the decision is reached. If either party requests, then the adjudicator must give reasons in the decision. If an adjudicator gives any reasons, they are to be read with the decision and may be used as a means of interpreting and understanding the decision and the reasons for that decision. Comments about the decision given by the adjudicator after delivering the decision are irrelevant.

In the absence of any directions about the time to comply, compliance must be immediate on delivery of the decision to the parties. The decision is binding and must be complied with until the dispute is finally determined by arbitration, legal proceedings or agreement.

The adjudicator is entitled to reasonable fees which the adjudicator may determine. The parties are jointly and severally liable for payment if the adjudicator makes no apportionment or if there is an outstanding balance. However, an adjudicator who does not produce a decision which is enforceable may not be entitled to a fee. Much depends on what it says in the adjudicator's terms of engagement.

211

The adjudicator will not be liable for anything done or omitted in carrying out the functions of an adjudicator unless the act or omission is in bad faith. Similar protection is also given to any employee or agent of the adjudicator. It is perhaps worth noting that, as an incorporated term of the contract, this paragraph is not binding on persons who are not parties to the contract.

The Scheme: Costs

Nothing in the Scheme allows the adjudicator to award the parties costs. This is in harmony with the philosophy of the Act, which does not encourage the parties to incur large amounts of costs in pursuing claims. In arbitration and litigation, by contrast, where costs are normally awarded against the losing party, the dispute can deteriorate into a fight about costs rather than the point at issue. That is because of the huge costs which can be incurred by each side. The adjudicator can be given power to award costs. The precise way in which this can be achieved has been the subject of some dispute.

212

23.7 Arbitration

General

Arbitration is relatively uncommon in the construction industry. Its place has been taken in large measure by adjudication undoubtedly because of the relative cheapness and undoubted speed of adjudication compered to arbitration. For that reason, it is unlikely that most architects will be involved in arbitration, especially when using the MW or MWD. Even if the contractor seeks arbitration, the employer will have to engage appropriate professional assistance to deal with it. Any involvement by the architect will only be as a witness. Therefore, this will be a brief overview. Arbitration can take place on any matter at any time. Arbitrators appointed under the JCT Agreement are given extremely wide powers. Their jurisdiction is to decide any dispute or difference of any kind whatsoever arising out of or in connection with the contract. The scope could scarcely be broader. The arbitrator's powers are to:

Article 7, clause 7.3, Schedule 1

- Rectify the contract so that it shows what the parties actually agreed;
- Direct the taking of measurements or the undertaking of whatever valuations the arbitrator thinks desirable to determine the respective rights;
- Ascertain and make an award of any sum which should have been included in a certificate;

Schedule 1,
paragraph 3

■ Open up, review and revise any certificate, opinion, decision, requirement or notice issued, given or made and to determine all matters in dispute as if no such certificate, opinion, decision, requirement or notice had been given.

Governing rules

Schedule 1,
paragraph 1
Schedule 1,
paragraph 6

1.7

The JCT 2016 edition of the Construction Industry Model Arbitration Rules (CIMAR) current at the contractual base date, are to govern the proceedings. The provisions of the Arbitration Act 1996 are expressly stated to apply to any arbitration under this agreement. That is to be the case no matter where the arbitration is conducted. Therefore, even if the project and the arbitration takes place in a foreign jurisdiction, the UK Act will apply provided that the parties contracted on MW or MWD and the contract provision that the law of England applies is not amended.

The following matters are excluded from arbitration:

■ Disputes about Value Added Tax, where legislation provides some other method of resolving the dispute.
■ Disputes under the Construction Industry Scheme, where legislation provides some other method of resolving the dispute.
■ The enforcement of any decision of an adjudicator.

Questions of law

213

Schedule 1,
paragraph 5

214

The contract records that the employer and contractor agree, in accordance with the relevant sections of the Arbitration Act, that either party may by proper notice to the other and to the arbitrator apply to the courts to decide any question of law arising in the course of the proceedings and appeal to the courts on any question of law arising out of an award. Clauses like this have been held by the courts to be effective.

A formal process

Arbitrations are conducted quite formally, like private legal proceedings – which is what they are. An arbitrator begins by inviting the parties to the 'preliminary meeting', but that does not mean a friendly discussion. It is a formal meeting to establish all the important criteria which need to be decided before the arbitration can proceed. A party should not go to a preliminary meeting without taking a fully briefed legal adviser experienced in arbitration.

The employer and contractor are free to agree who should be appointed, or should appoint, the arbitrator and they have freedom to agree important matters such as the form and timetable of the proceedings. This raises the possibility of a quicker procedure than would otherwise be the case in litigation and even matters such as the venue for any future hearing might be arranged to suit the convenience of the parties and their witnesses.

Hearings

If oral evidence and cross-examination is to be carried out, it is usually done at a hearing. Hearings, which are the private equivalent of a trial, are conducted in private, not in an open court.

Procedure

Schedule 1, paragraph 1

Arbitrations begun under the contract must be conducted subject to and in accordance with the JCT 2016 edition of CIMAR, current at the base date of the contract. If any amendments have been issued by JCT since that date, the parties may jointly agree to give written notice the arbitrator to conduct the reference according to the amended rules. CIMAR is a comprehensive body of rules, generally of admirable clarity. The whole document, at the time of writing, is available on www.jctcontracts.com. As might be expected JCT/CIMAR is very detailed. Among other things, they offer the parties a choice of three broad categories of procedure by which the proceedings will be conducted, as follows:

Short hearing procedure

Rule 7

This procedure is not usual. It limits the time available to the parties within which to deal with the matters in dispute orally in front of the arbitrator. Before the hearing, each party provides the arbitrator and each other with a written statement of their claim, defence and counterclaim (if any). All relevant documents and witness statements relied upon must be attached. There are some time scales. The arbitrator may carry out an inspection.

This procedure is particularly suited to issues which fairly easily can be decided by such an inspection of work, materials, plant and/or equipment or the like. The arbitrator must decide the issues and make an award within a month after concluding hearing the parties. Expert evidence is possible but costly and often unnecessary. Parties can agree to allow the arbitrator to use specialist expertise when reaching the decision.

This procedure with a hearing is ideally suited to many disputes which are relatively simple and provides for a quick award with minimum delay and associated cost.

Documents only procedure

Rule 8

This will rarely be used. It is not viable unless all the evidence is in the form of documents. Nevertheless, it can offer economies of time and cost. It is best suited where the sums in issue are relatively modest. The parties, in accordance with a timetable devised by the arbitrator, will serve on each other and on the arbitrator a written statement of case which will include an account of the relevant facts and opinions upon which reliance is placed and a statement of the precise remedy sought.

If either side is relying on witnesses of fact, the witness statements will be included with the statement of case. If the opinions of an expert or experts are required, they must also be given in writing and signed. There is a right of reply and if there is a counterclaim, the other party may reply to it.

Although referred to as 'documents only', the arbitrator may set aside up to a day to question the parties and/or their witnesses. The arbitrator must make a decision within a month or so of final exchanges and questioning, but the arbitrator may notify the parties that more time for the decision will be required.

Full procedure

Rule 9   If neither of the other options is considered suitable, CIMAR makes provision for the parties to conduct their cases in a similar way to normal High Court proceedings, giving the opportunity to hear and cross examine factual and expert witnesses. This is the most complex procedure. The parties must exchange formal statements. Sometimes multiple statements may be exchanged. Each submission must be detailed to enable the other party to answer each allegation made.

The arbitrator should give detailed directions concerning everything necessary for the proper conduct of the arbitration. Directions may also be given requiring disclosure of documents.

The appointment of an arbitrator

It is at the option of either party to begin arbitration proceedings. As a first step, one party must write to the other requesting them to agree to the appointment *Schedule 1,* of an arbitrator. Whoever does so, proceedings are formally commenced when *paragraph 2.1* the written notice is received. The notice must identify the dispute. It is good practice for the party seeking arbitration to insert the names of three prospective arbitrators. This saves time and often both parties can agree on one of the names. The arbitrator must have no relationship to either of the parties and no connection with any matter associated with the dispute.

It is important for the parties to make a real effort to agree on a suitable candidate rather than having one appointed whose skills and experience may be entirely unknown. If the parties cannot agree upon a suitable appointment within 14 days of a notice to agree or any extension to that period, either party *Schedule 1,* can apply to a third party to appoint an arbitrator. There is a list of appointors *paragraph 2.1* in the contract particulars. All but one should be deleted. The default provision *and Rule 2.3* is the President or a Vice-President of the Royal Institute of British Architects It will be necessary to complete special forms and to pay the relevant fee. Although the system of appointing an arbitrator varies, the aim is the same. The object is to appoint a person of integrity who is independent, having no existing relationships with either party or their professional advisers and who is impartial. It should go without saying that the arbitrator should have the necessary and appropriate technical and legal expertise. Claimants who have a dispute to refer and respondents receiving a notice to agree should waste no time in taking proper expert advice on how best to proceed.

If the arbitrator's appointment is made by agreement, it will not take effect until the appointed person has confirmed willingness to act, irrespective of whether terms have been agreed. If the appointment is the result of an application to the appointing body, it becomes effective, whether or not terms have *Rule 2.5* been agreed, when the appointment is made by the relevant body. There is no fixed scale of charges for arbitrator's services and fees ought to depend on their experience, expertise and often on the complexity of the dispute. Arbitrators usually require an initial deposit from the parties and, if there is to be a hearing, there will be a cancellation charge graded in accordance with the proximity of the cancellation to the start of the hearing. A cancellation means that it is difficult for the arbitrator to secure work at short notice to fill the void. In cases where the cancellation fee is substantial, due to proximity to the hearing date, it might be sensible to ask the arbitrator to account to the parties for activities during the hearing period.

After appointment, the arbitrator will consider which of the procedures summarised above appears to be most appropriate. The arbitrator must choose the format that will best avoid undue cost and delay. Therefore, parties must within 14 days after acceptance of the appointment is notified to the parties, provide the arbitrator with an outline of their disputes and of the sums in issue along with an indication of which procedure they consider best suited to them. After due consideration of all parties' views and unless a meeting is considered unnecessary, the arbitrator must, within 21 days of the date of acceptance, arrange a meeting (the preliminary meeting) which the parties or their representatives will attend to agree (if possible) or receive the arbitrator's decision upon everything necessary to enable the arbitration to proceed. It is obviously preferable for the parties to agree which procedure is to apply. If they cannot agree, the 'documents only' procedure will apply unless the arbitrator, after having considered all representations, decides that the full procedure will apply.

The parties are always free to conduct their own cases, but if disputes have reached the stage of formal proceedings it is usually better to engage experienced professionals to act for them.

Powers of the arbitrator

The 1996 Arbitration Act significantly broadened the arbitrator's powers beyond what was previously the case. For example, an arbitrator may:

- Order which documents or classes of documents should be disclosed between and produced by the parties;
- Order whether the strict rules of evidence shall apply;
- Decide the extent to which the arbitrator should take the initiative in ascertaining the facts and the law;
- Take legal or technical assistance or advice;
- Order security for costs;

- Give directions in relation to any property owned by or in the possession of any party to the proceedings which is the subject of the proceedings;
- Make more than one award at different times on different aspects of the matters to be determined;
- Award interest;
- Make an award on costs of the arbitration between the parties;
- Direct that the recoverable costs of the arbitration, or any part of the arbitral proceedings, are to be limited to a specified amount.

Figure 23.2 shows the outline of adjudication and arbitration in simple flow-chart form.

Third party procedure

One of the perceived advantages of litigation over arbitration is that claimants can take action against several defendants at the same time and any defendant can seek to join into the proceedings another party who may have liability. For example, if X sues Y, Y may sue Z as part of the same proceedings if Y thinks that Z is responsible for some or all of the dispute. This facility is not readily available in arbitration which usually takes place only between the parties to the contract.

Rules 2.6 and 2.7

However, where there are two or more related sets of proceedings on the same topic, but under different arbitration agreements, anyone who is charged with appointing an arbitrator must consider whether the same arbitrator should be appointed for both. In the absence of relevant grounds to do otherwise, the same arbitrator is to be appointed. If different appointors are involved they must consult one another. If one arbitrator is already appointed, that arbitrator must be considered for appointment to the other arbitrations.

This situation commonly occurs when there is an arbitration under the main contract and also between the contractor and a sub-contractor about the same issue, perhaps one of valuation or extension of time. It is also possible that there are two contracts between the same two parties and an issue arises in both which is essentially the same point. Usually, the same arbitrator ought to be appointed for that situation.

23.8 Legal proceedings (litigation)

Article 8

Article 7

The legal proceedings option simply provides that the English courts will have jurisdiction over any dispute or difference arising out of or in connection with the contract and will be determined by legal proceedings. Obviously, if the work is to be done in Northern Ireland or Scotland, the relevant courts will be substituted for the English courts. Parties wishing to adopt this procedure will delete the arbitration option. If neither option is deleted, legal proceedings is the default position.

23.9   Mediation

clause 7.1

The contract states that the parties should give 'serious consideration' to any request by the other to resolve any dispute by mediation. It is unclear why this clause has been included in the contract at all, because mediation depends on both parties agreeing. One assumes that this clause was inserted purely to remind the less sophisticated users of the form that mediation is a possibility. In general, there is little point in including as terms of a contract anything to be agreed. The whole point of a written contract is that it is evidence of what the parties have already agreed. To have a clause which effectively states: 'both parties should serious think about agreeing to do something else', is a waste of space.

In practice, mediation seems to be effective if the mediator is competent and if the parties really do want to compromise. Mediation can take whatever form the parties agree. But if one or both parties are determined to stick to their position come what may, mediation is a waste of time.

23.10   Common problems

If the adjudicator's decision is obviously wrong

It is not of course unusual for the losing party to think the adjudicator got the decision wrong, but sometimes the decision is very obviously wrong. That will not stop the winner taking advantage of it and, if the loser does not pay, the decision can be enforced by the court. The court will not be interested in whether the adjudicator came to the correct decision, but only whether the decision was made in the right way. For example, the court will want to be sure that the adjudicator had the jurisdiction to decide the dispute and that he or she answered the right question. It has been said that if the adjudicator answers the right question with the wrong answer, it will be enforced, but if the adjudicator answers the wrong question with the right answer it will not be enforced. That is because adjudication is all about getting a binding decision quickly. If the loser wishes, the matter can then be referred to arbitration or litigation (whichever is chosen in the contract). The arbitrator or judge is not reviewing the adjudicator's decision, but rather deciding the question afresh.

If the other party is not responding at all, will that stop the adjudication?

Occasionally, one of the parties to the contract will not respond to a notice of adjudication. Sometimes the party has moved address and cannot be found. Neither situation will stop the adjudication (or an arbitration or legal proceedings) going ahead. The adjudicator will make every reasonable effort to involve the missing party, but if all attempts fail, the adjudicator will proceed with only one party and come to a decision. If the decision is against the missing party it is likely that the winner will have difficulty recovery any money ordered to be paid, but that is a different problem.

Notes and references

Chapter 1

1   *Bowmer and Kirkland Ltd v Wilson Bowden Properties Ltd* (1996) 80 BLR 131.
2   Architects Act 1997.
3   S. 59 of the Finance Act 2004.
4   *Temloc Ltd v Erill Properties Ltd* (1987) 39 BLR 30.
5   *Pozzolanic Lytag Ltd v Bryan Hobson Associates* (1999) BLR 267.
6   *Brodie v Cardiff Corporation* (1919) AC 337.
7   This provision was added following the decision in *Co-operative Insurance Society v Henry Boot Scotland Ltd* (2002) EWHC 1270 (TCC).
8   See *Equitable Debenture Assets Corporation Ltd v William Moss* (1984) 2 Con LR 1; *Victoria University of Manchester v Hugh Wilson and Lewis Womersley and Pochin (Contractors) Ltd* (1984) 2 Con LR 43; *University Court of the University of Glasgow v William Whitfield and John Laing (Construction) Ltd* (1988) 42 BLR 66; *Edward Lindenberg v Joe Canning and Jerome Contracting Ltd* (1992) 29 Con LR 71; *Plant Construction plc v Clive Adams Associates and JMH Construction Services Ltd* (2000) BLR 137.
9   *English Industrial Estates Corporation Ltd v George Wimpey and Co Ltd* (1972) 7 BLR 122; *Gleeson v Hillingdon London Borough* (1970) 215 EG 165.
10   *Williams v Fitzmaurice* (1858) 157 ER 709.
11   *Douglas (R M) Construction Ltd v CED Building Services* (1985) 3 Con LR 124.
12   *London Borough of Merton v Stanley Hugh Leach Ltd* (1985) 32 BLR 51.
13   *Penwith District Council v V. P. Developments Ltd* (1999) EWHC Technology 231, *Cantrell and Another v Wright and Fuller Ltd* (2003) 91 Con LR 97.
14   Copyright, Designs and Patents Act 1988.
15   Clauses 1.4 and 1.6 give effect to sections 115 and 116 of the Housing Grants, Construction and Regeneration Act 1996 (as amended).
16   Contracts (Rights of Third Parties) Act 1999.
17   *Lazerbore v Morrison Biggs Wall* (1993) CILL 896.
18   *Serck Controls Ltd v Drake and Scull Engineering Ltd* (2000) 73 Con LR 100.
19   *Tameside Metropolitan Borough Council v Barlows Securities Group Services Ltd* (2001) BLR 113.

Chapter 2

20   David Chappell, Michael Cowlin, Michael Dunn; Building Law Encyclopaedia, (2009) Wiley-Blackwell, pp. 172–173.

21 The Copyright, Designs and Patents Act 1988.

22 *Bowmer and Kirkland Ltd v Wilson Bowden Properties Ltd* (1996) 80 BLR 131.

23 *Tweddle v Atkinson* (1861) 1 B and S 393.

24 *McGruther v Pitcher* (1904) 2 Ch 306; *Adler v Dickson* (1954) 3 All ER 788; *Scruttons Ltd v Midland Silicones Ltd* (1962) 1 All ER 1.

25 The Contracts (Rights of Third Parties) Act 1999.

26 *Fenice Investments Inc v Jerram Falkus Construction Ltd* (2009) 128 Con LR 124.

Chapter 3

27 In Northern Ireland, legislation to the same effect is the Construction Contracts (Northern Ireland) Order 1997 which was amended in line with the 2009 Act by the Construction Contracts (Amendment) Act (Northern Ireland) 2011 which came into force on 14 November 2012.

28 *Luxor (Eastbourne) Ltd v Cooper* (1941) 1 All ER 33.

29 *London Borough of Merton v Stanley Hugh Leach Ltd* (1985) 32 BLR 51.

30 *Symonds v Lloyd* (1859) 141 ER 622.

31 *McCutcheon v David McBrayne Ltd* (1964) 1 WLR 125.

32 *Hancock v B W Brazier (Anerley) Ltd* (1966) 2 All ER 901; *Young and Marten Ltd v McManus Childs Ltd* (1969) 9 BLR 77; *Test Valley Borough Council v Greater London Council* (1979) 13 BLR 63; *IBA v EMI and BICC* (1980) 14 BLR 1.

33 *Tameside Metropolitan Borough Council v Barlows Securities Group Services* Ltd (2001) BLR 113.

34 Section 32 of the Act.

35 *Gray and Others v T P Bennett and Son and Others* (1987) 43 BLR 63; *Sheldon and Others v R H M Outhwaite (Underwriting Agencies) Ltd* (1995) 2 All ER 558; *Cave v Robinson Jarvis and Rolf* (2003) 1 AC 384.

36 *British Steel Corporation v Cleveland Bridge Company* (1984) 1 All ER 504.

37 *Laserbore v Morrison Biggs Wall* (1993) CILL 896.

38 *Bouygues (UK) Ltd v Dahl-Jensen UK Ltd* (2000) BLR 522.

Chapter 4

39 Architects cannot rely on the Court of Appeal decision in *Pacific Associates Inc. v Baxter* (1988) 16 Con LR 90.

40 Under the rule in *Hedley Byrne and Partners v Heller and Co Ltd* (1963) 2 All ER 575.

41 *Richard Roberts Holdings Ltd v Douglas Smith Stimson and Partners* (1988) 22 Con LR 69.

42 *Holt v Payne Skillington* (1995) 77 BLR 51. Or of course they may be less: *Aiken v Stewart Wrightson Agency* (1995) 1 WLR 1281. Where contractual and tortious duties exist side by side, the test for recoverability of damages is the contractual one: *Wellesley Partners LLP v Withers LLP* (2015) 163 Con LR 53.

43 *Henderson v Merrett Syndicates* (1994) 69 BLR 29; *Conway v Crowe Kelsey and Partners* (1994) 39 Con LR 1.

44 *J. Jarvis and Sons Ltd v Castle Wharf Developments and Others* (2001) Lloyds Rep 308.

45 *Croudace Ltd v London Borough of Lambeth* (1986) 6 Con LR 72.

46 *Penwith District Council v V P Developments* (1999) EWHC Technology 231.

47 *Bolam v Friern Hospital Management Committee* (1957) 2 All ER 118.

48 *Wimpey Construction UK Ltd v Poole* (1984) 27 BLR 58.

49 *De Freitas v O'Brien* (1995) PIQR P281.

50 *B L Holdings v Robert J Wood and Partners* (1979) 12 BLR 1.

51 *Sutcliffe v Thackrah* (1974) 1 All ER 319.

52 *Michael Sallis and Co. Ltd v Calil and W F Newman and Associates* (1987) 12 Con LR 68. The decision was called into question by the Court of Appeal in *Pacific Associates Inc v Baxter* (1988) 16 Con LR 90, a case which concerned a firm of consulting engineers. The grounds for the *Pacific Associates* decision involved the existence of an arbitration clause in the main contract and an exclusion of liability clause. The existence of an exclusion of liability clause is not relevant to the question of whether a duty of care exists. The arguments put forward in *Sallis* may yet be revived.

53 *Perini Corporation v Commonwealth of Australia* (1969) 12 BLR 82.

54 *Rees and Kirby Ltd v Swansea Corporation* (1983) 25 BLR 129.

55 *Croudace Ltd v London Borough of Lambeth* (1986) 6 Con LR 72.

56 *London County Council v Vitamins Ltd* (1955) 2 All ER 229.

57 *Scheidebouw BV v St James Homes (Grosvenor Dock) Ltd* (2006) 105 Con LR 90.

58 *Token Construction Co Ltd v Charlton Estates Ltd* (1973) 1 BLR 48.

59 *Bowmer and Kirkland Ltd v Wilson Bowden Properties Ltd* (1996) 80 BLR 131

60 *London Borough of Merton v Stanley Hugh Leach Ltd* (1985) 32 BLR 51.

61 *Bernuth Lines Ltd v High Seas Shipping Ltd* (2005) EWHC 3020 (Comm); *Construction Partnership UK Ltd v Leek Developments* (2006) EWHC B8 (TCC).

62 *Golden Ocean Group Ltd v Salgaocar Mining Industries PVT Ltd and Another* (2012) 3 All ER 842.

63 *Proton Energy Group SA v Orlen Lietuva* (2013) 150 Con LR 72.

Chapter 5

64 *Thompson (M C) v Clive Alexander and Partners* (1992) 8 Const LJ 199.

65 For example, *English Industrial Estates Corporation Ltd v George Wimpey and Co Ltd* (1972) 7 BLR 122; *Fenice Investments Inc v Jerome Falkus Construction Ltd* (2009) 128 Con LR 124.

66 *Glenlion Construction Ltd v The Guinness Trust* (1987) 11 Con LR 126, on slightly different wording ('on or before the date for completion').

67 This is the same inconsistency which troubled the court in *Greater London Council v Cleveland Bridge and Engineering Co Ltd* (1984) 8 Con LR 30 and whose judgment was wholeheartedly endorsed by the Court of Appeal (1986) 8 Con LR 44. There, the court took the view that, although a failure to proceed with due diligence allowed the employer to discharge the contractor, it did not amount to a breach of contract.

68 *West Faulkner v London Borough of Newham* (1994) 42 Con LR 144.

69 *Hampshire County Council v Stanley Hugh Leach Ltd* (1990) 8 Const LJ 174.

70 *Bluewater Energy Services BV v Mercon Steel Structures BV and Others* (2014) 155 Con LR 85.

71   *Rotherham Metropolitan Borough Council v Frank Haslam Milan and Co Ltd and Ano* (1996) EGCS 59.

72   *Perry v Tendring District Council* (1985) 3 Con LR 74.

73   *Brazier v Skipton Rock Co Ltd* (1962) 1 All ER 955.

74   This provision puts the relevant provision in the Housing Grants, Construction and Regeneration Act 1996 (as amended) into effect.

75   Section 112(4) of the Housing Grants, Construction and Regeneration Act 1996 (as amended).

76   Housing Grants, Construction and Regeneration Act 1996 (as amended).

77   The Scheme for Construction Contracts (England and Wales) Regulations 1998 (as amended).

Chapter 6

78   *Luxor (Eastbourne) Ltd v Cooper* (1941) 1 All ER 33.

79   *Cory v City of London Corporation* (1951) 2 All ER 85.

80   *Rapid Building Co Ltd v Ealing Family Housing Association Ltd* (1985) 1 Con LR 1.

81   *C J Elvin Building Services Ltd v Peter and Alexa Noble* (2003) EWHC 837 (TCC) supports the view that repeated failures and sometimes even one failure to pay may amount to repudiation.

82   These have been made necessary by the provisions of the Housing Grants, Construction and Regeneration Act 1996 (as amended).

83   *Rupert Morgan Building Services (LLC) Ltd v David Jervis and Harriett Jervis* (2004) BLR 18.

84   *Wates Construction Ltd v Franthom Property Ltd* (1991) 53 BLR 23.

85   *London Borough of Lewisham v Shepherd Hill Engineering*, 30 July 2001 unreported.

86   *J F Finnegan Ltd v Ford Seller Morris Developments Ltd (No. 1)* (1991) 25 Con LR 89.

87   *Scheidebouw BV v St James Homes (Grosvenor Dock) Ltd* (2006) 105 Con LR 90.

Chapter 7

88   *Picardi v Mr and Mrs Cuniberti* (2003) 19 Const LJ 350.

89   *John Laing Developments Ltd v County and District Properties Ltd* (1982) 23 BLR 1.

90   *Burden (R B) v Swansea Corporation* (1957) 3 All ER 243.

91   *Sutcliffe v Thackrah* (1974) 1 All ER 319.

92   *Clusky (trading as Damian Construction) v Chamberlain* (1995) April BLM 6.

Chapter 8

93   *Simplex Concrete Piles Ltd v Borough of St Pancras* (1958) 14 BLR 80.

94   This is called 'vicarious liability' – the liability of one party for the wrongs committed by another. The most common example is the liability of an employer for the actions of an employee: *Kensington and Chelsea and Westminster Area Health Authority v Wettern Composites and Others* (1984) 1 Con LR 144; *Gray and Others v T P Bennett and Son* (1987) 43 BLR 63.

Chapter 9

95   The effectiveness of this kind of clause was considered by the House of Lords in *Linden Gardens v Lenesta Sludge Disposals* (1993) 9 Const LJ 322 and *St Martin's Property Corporation v Sir Robert McAlpine and Sons* (1993) 9 Const LJ 322.

96   *Moresk Cleaners v Hicks* (1966) 4 BLR 50.

97   *Vonlynn Holdings Ltd v Patrick Flaherty Contracts Ltd*, 26 January 1988, unreported; *AMEC Building Contracts Ltd v Cadmus Investments Co Ltd* (1997) 13 Const LJ 50.

Chapter 10

98   *London Borough of Merton v Stanley Hugh Leach Ltd* (1985) 32 BLR 51.

99   That position is not so clear cut since the decision in *Murphy v Brentwood District Council* (1990) 50 BLR 1.

Chapter 11

100   *Pozzolanic Lytag v Bryan Hobson Associates* (1999) BLR 267.

101   The very instructive case of *National Trust for Places of Historic Interest or Natural Beauty v Haden Young* (1994) 72 BLR 1 provided a clear and sensible explanation of the way the predecessor of this clause should work in connection with clause 5.4B. The case concerned an earlier version of the contract, and it is now clear that the contractor gives no indemnity under clause 5.2, nor does it insure, in respect of the Works, or existing buildings to which the work is being carried out or the contents of such existing buildings. If damage is caused to the Works through the contractor's fault, the contractor may be liable, because its obligation is to carry out and complete the Works in accordance with the contract documents. The crucial question was whether the predecessors of clauses 5.2 and 5.4B provided a scheme whereby loss to be insured by the employer fell upon the employer under a different wording; the trial judge emphatically said 'No' and the Court of Appeal agreed with him.

　　　　Essentially, clause 5.2 makes the contractor liable for damage to property, except the Works which are the subject of the building contract and property insured under clause 5.4B, to the extent that the damage is due to the contractor's or any sub-contractor's negligence or default. The contractor will be liable to a partial extent if it is partially at fault and damage is caused to surrounding buildings, passing vehicles or an existing building to which the Works are being carried out. The Works are expressly excepted, but if damage is caused to the Works themselves through the contractor's fault, the contractor may be liable, because its obligation is to carry out and complete the Works in accordance with the contract documents. The contractor is obliged to carry appropriate insurance. Effectively, the employer and contractor agree that if there is damage to the existing structure or contents caused by specified perils, this clause would deal with the situation even if due to the contractor's negligence.

102   *London Borough of Barking and Dagenham v Stamford Asphalt Co* (1997) 82 BLR 25.

103   *Gold v Patman and Fotheringham Ltd* (1958) 2 All ER 497.

104   A full definition is to be found in clause 1.1 Definitions.

105   *Horbury Building Systems Ltd v Hampden Insurance NV* (2004) EWCA Civ 418.

106   *TFW Printers Ltd v Interserve Project Services Ltd* (2006) EWCA Civ 875.

107   This is the interpretation given to 'forthwith' in *London Borough of Hillingdon v Cutler* (1967) 2 All ER 361. A stricter interpretation was given as 'immediately or as quickly as possible by what is currently regarded as conventional and universally available methods of business communication' by a Scottish court in *St Andrews Bay Development Ltd v HBG Management Ltd and Mrs Janey Milligan* (2003) Scot CS 103.

Chapter 12

108   *Freeman v Hensler* (1900) 64 JP 260.

109   *Hounslow Borough Council v Twickenham Garden Developments* (1971) 7 BLR 81.

110   In *Greater London Council v Cleveland Bridge and Engineering Co. Ltd* (1986) 8 Con LR 30, the court was of the opinion that a requirement for the contractor to proceed with due diligence and expedition must be interpreted in the light of other requirements as to time. The contractor may, therefore, be able to argue that it is proceeding regularly and diligently provided it can meet the completion date and it is a moot point whether it could be said to be suspending something it had not yet begun. In any case, it might well argue that it had reasonable cause if, in fact, the contract period was very generous. 'Regularly and diligently' has now been comprehensively defined by the Court of Appeal: *West Faulkner v London Borough of Newham* (1994) 42 Con LR 144 (See Chapter 5).

111   *Rapid Building Co Ltd v Ealing Family Housing Association Ltd* (1985) 1 Con LR 1.

112   *Hounslow Borough Council v Twickenham Garden Developments* (1971) 7 BLR 81.

Chapter 13

113   *Percy Bilton Ltd v Greater London Council* (1982) 20 BLR 1.

114   *Wells v Army and Navy Co-operative Society Ltd* (1902) 86 LT 764.

115   *Percy Bilton Ltd v Greater London Council* (1982) 20 BLR 1.

116   The contract uses the word 'thereupon' which means 'immediately after that.'

117   *Balfour Beatty Construction Ltd v London Borough of Lambeth* (2002) BLR 288.

118   *Royal Brompton Hospital v Frederick Alexander Hammond (No. 7)* (2001) 76 Con LR 148.

119   This clarification was added in the wake of *Scott Lithgow v Secretary of State for Defence* (1989) 45 BLR 1, when the court surprisingly held that, in some circumstances, sub-contractors may not be under the control of the contractor.

120   In *Wells v Army and Navy Co-operative Society Ltd* (1902) 86 LT 764, a very similar phrase, 'other causes beyond the contractor's control' was held not to extend to delays caused by the employer or his architect.

121   *London Borough of Merton v Stanley Hugh Leach Ltd* (1985) 32 BLR 51.

122   *Walter Lawrence and Son Ltd v Commercial Union Properties* (1984) 4 Con LR 37.

Chapter 14

123   *Cavendish Square Holding BV v Talal El Makdessi* (2015) 162 Con LR 1.

124   *BFI Group of Companies Ltd v DCB Integrated Systems Ltd* (1987) CILL 348, a case on MW 80 holding that liquidated damages are payable even if there is no loss. *Impresa Castelli SpA v Cola Holdings Ltd* (2002) 87 Con LR 123 is a later case to much the same effect.

125   *Phillips Hong Kong Ltd v The Attorney General of Hong Kong* (1993) 9 Const LJ 202.

126   *North Sea Ventilation Ltd v Consafe Engineering (UK) Ltd* 20 July 2004, unreported.

127   *Temloc Ltd v Erill Properties Ltd* (1987) 39 BLR 30.

128   *Kemp v Rose* (1858) 1 Giff 258.

129   *London Borough of Lewisham v Shepherd Hill Civil Engineering* 30 July 2001, unreported.

130   *British Glanzstoff Manufacturing Co. Ltd v General Accident Fire and Life Assurance Corporation Ltd* (1913) AC 143.

Chapter 15

131   *Holland Hannen and Cubitt (Northern) Ltd v Welsh Health Technical Services Organisation* (1981) 18 BLR 80.

Chapter 16

132   *London Borough of Hillingdon v Cutler* (1967) 2 ALL ER 361.

133   *Bowmer and Kirkland Ltd v Wilson Bowden Properties Ltd* (1996) 80 BLR 131; *Redheugh Construction Ltd v Coyne Contracting Ltd and British Columbia Building Corporation* (1997) 29 CLR (2d) 39-46; *Ministry of Defence v Scott Wilson Kirkpatrick and Dean and Dyball Construction* (2000) BLR 20 reach broadly similar conclusions on this point. However, a contrary view was taken in *W S Harvey (Decorators) Ltd v H L Smith Construction Ltd* 5 March 1997 unreported.

134   *Rhuddlan Borough Council v Fairclough Building Ltd* (1985) 3 Con LR 38.

135   Otherwise the work could be postponed several times during its course, at great expense to the contractor and it would have no recourse.

136   *Bath and North East Somerset District Council v Mowlem* (2004) EWCA Civ 115.

137   *John Jarvis v Rockdale Housing Association Ltd* (1986) 10 Con LR 51.

Chapter 17

138   *Abbey Developments Ltd v PP Brickwork Ltd* (2003) EWHC 1987 (TCC).

139   *Greater London Council v Cleveland Bridge and Engineering Company* (1984) 8 Con LR 30, upheld on appeal (1986) 8 Con LR 44.

140   Oxford English Dictionary.

Chapter 18

141   *Hoenig v Isaacs* (1952) 2 All ER 176.

142   *Traditional Structures Ltd v H W Construction Ltd* (2010) EWHC 1530 (TCC).

143   In the case of 'construction contracts' covered by the Housing Grants, Construction and Regeneration Act 1996 (as amended), section 109(2) makes clear that the parties are free to agree the amounts of payments and the intervals or circumstances in which they become due. This allows the parties to opt for stage payments. If the employer is a 'residential occupier' (section 106) which will often be the case with these contracts, the Act does not apply at all and the parties are in any event free to make their own arrangements.

144   *Sutcliffe v Chippendale and Edmondson* (1971) 18 BLR 149.

145   *Dawber Williamson Roofing Co Ltd v Humberside County Council* (1979) 14 BLR 70.

146   *Holland v Hodgson* (1872) Law Reports 7 Common Pleas 328; *Reynolds v Ashby* (1904) AC 466.

147   *Melluish v BMI (No. 3) Ltd* (1966) AC 454.

148   *Penwith District council v V P Developments* (1999) EWHC Technology 231; *Cantrell and Another v Wright and Fuller* (2003) 91 Con LR 97.

149   The Late Payment of Commercial Debts (Interest) Act 1998.

150   *Wilson and Sharp Investments Ltd v Harbour View Developments Ltd* (2015) 162 Con LR 154.

151   Limitation Act 1980.

Chapter 19

152   *Walter Lilly and Co. Ltd v Mackay and Another* (2012) 143 Con LR 79.

153   *Westminster Corporation v J Jarvis and Sons* (1970) 1 All ER 943; *Nevill (H. W.) (Sunblest) Ltd v Wm Press and Son Ltd* (1981) 20 BLR 78.

154   *McGlynn v Waltham Contractors Ltd and Others (No. 3)* (2007) 111 Con LR 1; *Mul v Hutton Construction Ltd* (2014) 154 Con LR 159.

155   *BFI Group of Companies Ltd v DCB Integrated Systems Ltd* (1987) CILL 348; *Impresa Castelli SpA v Cola Holdings Ltd* (2002) 87 Con LR 123.

156   *Hadley v Kemp* (1999) EMLR 586.

Chapter 20

157   *Plant Construction plc v Clive Adams Associates and JMH Construction Services Ltd* (2000) BLR 137.

158   Case law, e.g, *Cullen v Butler* (1816) 5 M and S 461; *Harrison v Blackburn* (1864) 17 CB(NS) 678; *Wells v Army and Navy Co-operative Society Ltd* (1902) 86 LT 764 and many other cases, has established that the phrase 'other faults' is likely to be interpreted *ejusdem generis* with defects and shrinkages; that is to say, faults of the same kind.

159   This approach was noted with approval in *William Tomkinson v Parochial Church Council of St Michael* (1990) 6 Const LJ 319.

160 *London and S W Railway v Flower* (1875) 1 CPD 77.

161 *Mul v Hutton Construction Ltd* (2014) 154 Con LR 159.

162 *Pearce and High Ltd v John P Baxter and Mrs A Baxter* (1999) BLR 101.

163 *William Tomkinson v Parochial Church Council of St Michael* (1990) 6 Const LJ 319; *Pearce and High Ltd v John P Baxter and Mrs A Baxter* (1999) BLR 101.

164 Limitation Act 1980.

165 *Charnock v Liverpool Corporation* (1968) 1 WLR 1498; In *Hick v Raymond and Reid* (1893) AC 22, Lord Watson said at page 32: 'When the language of a contract does not expressly, or by necessary implication, fix any time for the performance of a contractual obligation, the law implies that it shall be performed in a reasonable time. The rule is of general application…'

Chapter 21

166 *Photo Production Ltd v Securicor Transport Ltd* (1980) AC 827.

167 *Woodar Investment Development Ltd v Wimpey Construction UK Ltd* (1980) 1 All ER 571.

168 *Hill (J M) and Sons Ltd v London Borough of Camden* (1980) 18 BLR 31.

169 *John Jarvis Ltd v Rockdale Housing Association Ltd* (1986) 10 Con LR 51.

170 Chappell, Cowlin and Dunn, *Building Law Encyclopaedia*, (2009), Wiley-Blackwell.

171 *Leander Construction Ltd v Mulalley and Co Ltd* (2011) EWHC 3449 (TCC).

172 The case of *West Faulkner v London Borough of Newham*, (1994) 42 Con LR 144, gives excellent guidance on the topic.

173 *John Mowlem and Co. v Eagle Star and Others* (1995) 44 Con LR 134.

174 *Hill (J M) and Sons Ltd v London Borough of Camden* (1980) 18 BLR 31.

175 *Robin Ellis Ltd v Vinexsa International Ltd* (2003) BLR 373.

176 *Wilson and Sharp Investments Ltd v Harbour View Developments Ltd* (2015) 162 Con LR 154.

177 *Emson Eastern (in Receivership) v EME Developments* (1991) 55 BLR 114.

178 *Argyropoulos and Pappa v Chain Compania Naviera SA* (1990) 7-CLD-05-01.

Chapter 22

179 *Haulfryn Estate Co Ltd v Leonard J Multon and Partners and Frontwide Ltd*, 4 April 1990, unreported.

180 The judge in another court has stated that he would require the clearest possible contractual condition before finding the contractor liable for a design fault: *John Mowlem and Co. Ltd v British Insulated Callenders Pension Trust Ltd* (1977) 3 Con LR 64.

181 *Moresk Cleaners Ltd v Hicks* (1966) 4 BLR 50.

182 *Viking Grain Storage Ltd v T H White Installations Ltd* (1985) 3 Con LR 52.

183 *Co-operative Insurance Society Ltd v Henry Boot Scotland Ltd* (2002) EWHC 1270 (TCC).

184 If the contractor simply went ahead and corrected the divergence in whatever way took his fancy, it would be a breach of contract, because the contractor would have constructed something which was not in the contract documents or in an architect's instruction. The only way to prevent it being a breach of contract is for the architect to approve it first. The approval is best issued as an architect's instruction, being careful not to actually instruct but to indicate approval of the contractor's proposal.

Chapter 23

185 *Herschel Engineering Ltd v Breen Properties Ltd* (2000) BLR 272.

186 *Connex South Eastern Ltd v M. J. Building Services plc* (2005) 2 All ER 871.

187 *Bouygues United Kingdom Ltd v Dahl-Jensen UK Ltd* (2000) BLR 522.

188 *City Inn Ltd v Shepherd Construction Ltd* (2003) BLR 468 CA; (2000) SLT 781.

189 s.9 of the Arbitration Act 1996.

190 *Fiona Trust and Holding Corporation and Others v Privalov and Others* (2007) 114 Con LR 69 HL followed in *Air Design (Kent) Limited v Deerglen (Jersey) Limited* (2008) EWHC 3047 (TCC).

191 *McAlpine PPS Pipeline Systems Joint Venture v Transco plc* (2004) All ER (D) 145.

192 *Cantillon Ltd v Urvasco Ltd* (2008) 117 Con LR 1.

193 *Air Design (Kent) Ltd v Deerglen (Jersey) Ltd* (2008) EWHC 3047 (TCC); *Amec Group Ltd v Thames Water Utilities Ltd* (2010) EWHC 419 (TCC).

194 *Jerome Engineering Ltd v Lloyd Morris Electrical Ltd* (2002) CILL 1827.

195 *Workspace Management Ltd v YJL London Ltd* (2009) EWHC 2017 (TCC).

196 *Holt Insulation Ltd v Colt International Ltd*, 23 July 2001, unreported.

197 *Karl Construction (Scotland) Ltd v Sweeney Civil Engineering (Scotland) Ltd* (2002) 18 Const LJ 55; *Sindall Ltd v Solland* (2001) 80 Con LR 152.

198 *Pilon Ltd v Breyer Group PLC* (2009) EWHC 837 (TCC).

199 *R Durtnell and Sons Ltd v Kaduna Ltd* (2003) BLR 225.

200 *Linnett v Halliwells LLP* (2009) 123 Con LR 104.

201 *Hart Investments Ltd v Fidler and Another* (2007) BLR 30.

202 *P T Building Services Ltd v ROK Eurobuild Ltd* (2008) EWHC 3434 (TCC); *Linnett v Halliwells LP* (2009) 123 Con LR 104.

203 *Aveat Heating Ltd v Jerram Falkus Construction Ltd* (2007) EWHC 131 (TCC).

204 *Cubitt Building and Interiors Ltd v Fleetglade Ltd* (2006) 110 Con LR 36.

205 *Barnes and Elliott Ltd v Taylor Woodrow Holding Ltd* (2004) BLR 111.

206 *C and B Concept Design Ltd v Isobars Ltd* (2002) BLR 93.

207 *R v Greater Birmingham Appeal Tribunal ex parte Simper* (1973) 2 All ER 461.

208 *Balfour Beatty Construction Ltd v London Borough of Lambeth* (2002) BLR 288.

209 *Ritchie Brothers (PWC) Ltd v David Philip (Commercials) Ltd* (2005) CSIH 32; *Hart Investments Ltd v Fidler and Another* (2007) BLR 30; *Mott MacDonald Ltd v London and Regional Properties Ltd* (2007) 113 Con LR 33.

210 *Joinery Plus Ltd (in administration) v Laing Ltd* (2003) 19 Const LJ T47.

211 *Systech International Ltd v P C Harrington Contractors Ltd* (2012) 145 Con LR 1.

212 See paragraph 25 of the Scheme (as amended) and section 108A of the Housing Grants, Construction and Regeneration Act 1996 (as amended).

213 Sections 45(2)(a) and 69(2)(a).

214 *Vascroft (Contractors) Ltd v Seeboard plc* (1996) 52 Con LR 1.

Table of cases

Clause number index to text

The JCT Minor Works Building Contracts 2016, Fifth Edition. David Chappell.
© 2018 John Wiley & Sons Ltd. Published 2018 by John Wiley & Sons Ltd.

Subject index